Math Mammoth Square Roots and the Pythagorean Theorem

By Maria Miller

ISBN 978-1-954358-74-4

2023 EDITION

Contents

Introduction 5

Square Roots 7

Irrational Numbers 11

Cube Roots and Approximations of Irrational Numbers 15

Fractions to Decimals 19

Decimals to Fractions 21

Square and Cube Roots as Solutions to Equations 24

More Equations that Involve Roots 27

The Pythagorean Theorem 30

Applications of the Pythagorean Theorem 1 35

A Proof of the Pythagorean Theorem and of Its Converse 38

Applications of the Pythagorean Theorem 2 42

Distance Between Points 46

Review 49

Answers 55

More from Math Mammoth 77

Introduction

Math Mammoth Square Roots & The Pythagorean Theorem is a relatively short worktext focusing on irrational numbers, square roots, and the Pythagorean Theorem and its applications.

First, students learn about taking a square root as the opposite operation to squaring a number. They learn about irrational numbers, and how to find approximations to square roots both with a calculator and with a guess-and-check method. Students also practice placing irrational numbers on the number line, using mental math to find their approximate location.

Next, the book has a review lesson on how to convert fractions to decimals. The following lesson has to do with writing decimals as fractions, and teaches a method for converting repeating decimals to fractions.

Then it is time to learn to solve simple equations that involve taking a square or cube root, over the course of two lessons. After learning to solve such equations, students are now fully ready to study the Pythagorean Theorem and apply it.

The Pythagorean Theorem is introduced in the lesson by that name. Students learn to verify that a triangle is a right triangle by checking whether it fulfills the Pythagorean Theorem. They apply their knowledge about square roots and solving equations to solve for an unknown side in a right triangle when two of the sides are given.

Next, students solve a variety of geometric and real-life problems that require the Pythagorean Theorem. This theorem is extremely important in many practical situations. Students should show their work for these word problems to include the equation that results from applying the Pythagorean Theorem to the problem and its solution.

There are literally hundreds of proofs for the Pythagorean Theorem. In this book, we present one easy proof based on geometry (not algebra). As an exercise, students are asked to supply the steps of reasoning to another geometric proof of the theorem. Students also study a proof for the converse of the theorem, which says that if the sides of a triangle fulfill the equation $a^2 + b^2 = c^2$ then the triangle is a right triangle.

Our last topic is distance between points in the coordinate grid, as this is another simple application of the Pythagorean Theorem.

You can find matching videos for topics in this chapter at https://www.mathmammoth.com/videos/ (grade 7).

I wish you success with teaching math!

Maria Miller

Helpful Resources on the Internet

We have compiled a list of Internet resources that match the topics in this book. This list of links includes web pages that offer:

- **online practice** for concepts;
- online **games**, or occasionally, printable games;
- **animations** and interactive **illustrations** of math concepts;
- **articles** that teach a math concept.

We heartily recommend you take a look at the list. Many of our customers love using these resources to supplement the bookwork. You can use the resources as you see fit for extra practice, to illustrate a concept better, and even just for some fun. Enjoy!

https://l.mathmammoth.com/blue/pythagorean

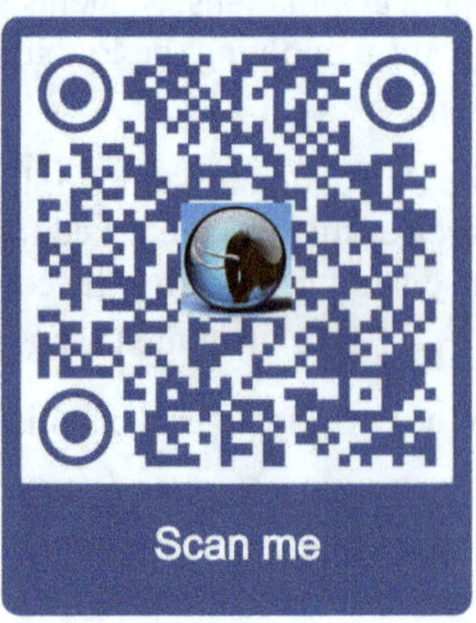

Square Roots

The **square** of a number is that number multiplied by itself. For example, six squared $= 6^2 = 6 \cdot 6 = 36$. (Recall that the square of 6 tells us the area of a square with sides 6 units long.)

Taking a **square root** is the opposite operation to squaring: the square root of 36 is the number that when squared, gives you 36.

There are actually two such numbers: 6 and −6. The positive one, 6, is **the principal square root** of 36. We use the "$\sqrt{\ }$" symbol (called the "radical sign" or "radix") to signify the principal square root of a number. For example, $\sqrt{25} = 5$ because $5^2 = 25$.

> The words "radish" and "radical" both come from the Latin word *radix*, meaning **root**.

Taking a square root allows us to find the side length of a square when its area is given.

Here is a way to remember what a square root is. In the picture on the right, the area of a square is written inside the square and the length of the side is written to the side:

49 7

Now, imagine the square is a radical sign that "houses" the number for the area:

To find the (principal) **square root of a number, think of a square with that area, and find the side length of that square.**

$\sqrt{49} = 7$

1. Find the (principal) square roots.

a. $\sqrt{100}$	b. $\sqrt{64}$	c. $\sqrt{4}$	d. $\sqrt{0}$
e. $\sqrt{81}$	f. $\sqrt{144}$	g. $\sqrt{1}$	h. $\sqrt{10{,}000}$

2. It is especially easy to find square roots of numbers that are **perfect squares:** numbers we get by squaring whole numbers. For example, 49 is a perfect square because it is 7^2. Fill in the list of perfect squares from 1^2 to 20^2 at the right:

x	x^2
1	1
2	4
3	9
4	_____
_____	_____
_____	_____
_____	49
8	_____
9	_____
_____	_____

x	x^2
11	_____
12	_____
13	_____
14	_____
15	_____
_____	256
_____	289
_____	324
_____	_____
_____	_____

3. Find the square roots of these perfect squares.

a. $\sqrt{169}$ b. $\sqrt{900}$

c. $\sqrt{225}$ d. $\sqrt{121}$

e. $\sqrt{441}$ f. $\sqrt{8{,}100}$

g. $\sqrt{324}$ h. $\sqrt{400}$

i. $\sqrt{6{,}400}$ j. $\sqrt{25{,}600}$

k. $\sqrt{16{,}900}$ l. $\sqrt{1{,}000{,}000}$

Most whole numbers are *not* perfect squares, and their square roots are unending decimals. (In fact, their square roots are **irrational numbers**, which means they cannot be written as a fraction, and their decimal expansions are unending decimals without any repeating patterns in the digits.)

We can handle that situation in at least three ways:

1. We can find an approximate value of such square roots **with a calculator**, rounding the answer to a reasonable accuracy. This is necessary if we're dealing with a real-life application.
2. We can find an approximate value using **guess and check**, and decimal multiplication. For example, we know that the value of $\sqrt{17}$ will be between 4 and 5 (since $\sqrt{16} = 4$ and $\sqrt{25} = 5$). We can also tell that it will be closer to 4 than 5, since 17 is very close to 16. So, we could guess that it is 4.1, square that, and based on the result, refine our guess.
3. We can **indicate such values using the square root symbol**, and not find a decimal approximation. For example, the side of a square with an area of 2 square units is $\sqrt{2}$ units. This is the preferred way in pure mathematics, and any time you want to convey an accurate value.

4. Between which two consecutive whole numbers do the following square roots lie? Do not use a calculator. Tell also which of those whole numbers the root is closer to.

a. $\sqrt{5}$ **b.** $\sqrt{24}$ **c.** $\sqrt{47}$ **d.** $\sqrt{83}$

5. Tell the side of the square (exact value) when its area is given. Indicate the side length using the square root symbol, if the area is not a perfect square. Note: u^2 signifies square units, and u signifies a unit.

a. area = 25 u^2 side = ________	**b.** area = 1,600 u^2 side = ________	**c.** area = 5 u^2 side = ________	**d.** area = 11 u^2 side = ________

6. **a.** What is the area of a square, if its side measures $\sqrt{8}$ units?

b. What is the value of $(\sqrt{7})^2$?

c. What is the side of a square with an area of 130 square meters? Give an exact value.

Example 1. Since $0.5 \cdot 0.5 = 0.25$, then $\sqrt{0.25} = 0.5$.

Example 2. Since $\frac{2}{3} \cdot \frac{2}{3} = \frac{4}{9}$, then $\sqrt{\frac{4}{9}} = \frac{2}{3}$.

7. Find the square roots.

a. $\sqrt{0.16}$ **b.** $\sqrt{0.01}$ **c.** $\sqrt{1.21}$

d. $\sqrt{\frac{16}{25}}$ **e.** $\sqrt{\frac{100}{9}}$ **f.** $\sqrt{\frac{49}{36}}$

Example 3. The area of a square is 42.5 m^2. What is the side of the square?

From a calculator, $\sqrt{42.5} \approx 6.5192024052026487145829715574292$. Even this long decimal is not giving us all the decimal digits! In reality, the number would continue in an unending manner, without any patterns in the decimal digits (it is an *irrational* number).

The area was given to three significant digits, so we will do the same here, and use three significant digits in our answer. The side measures 6.52 meters.

Note: On some calculators, you first push the square root button, then the number of which you are taking the square root. On others, you first enter the number and then push the square root button. Find out which way your calculator works.

8. Find the value of these square roots with a calculator, to three decimal digits.

a. $\sqrt{70}$	**b.** $\sqrt{3}$	**c.** $\sqrt{1{,}450}$
d. $\sqrt{0.45}$	**e.** $\sqrt{\frac{5}{6}}$	**f.** $\sqrt{\frac{31}{7}}$

The radical sign acts as a grouping symbol: it is as if there were parentheses around the expression under the square root. In other words, $\sqrt{15 + 10}$ means $\sqrt{(15 + 10)}$.

Example 4. Simplify $\sqrt{5 \cdot (70 + 10)}$.

We simplify the expression under the square root first, and take the square root last:

$$\sqrt{5 \cdot (70 + 10)} = \sqrt{5 \cdot 80} = \sqrt{400} = 20$$

9. Calculate. Do not use a calculator.

a. $\sqrt{9 + 16}$	**b.** $\sqrt{11 \cdot 11}$	**c.** $\sqrt{2 \cdot (41 - 9)}$
d. $\sqrt{225 - 9^2}$	**e.** $\sqrt{10^2 - 8^2}$	**f.** $\sqrt{13^2 - 12^2}$

10. Find the value of these expressions to three decimal digits with a calculator. Note: if your calculator doesn't automatically follow the order of operations, you need to use parentheses when entering the expressions. Another option is to write the intermediate results down or load them into the calculator's memory.

a. $\sqrt{5.6^2 - 2.1^2}$	**b.** $\sqrt{45.7^2 + 38.12^2}$

Use these exercises for more practice.

11. **a.** What is the area of a square if its side measures $\sqrt{1{,}600}$ cm?

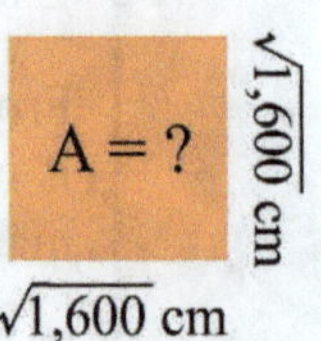

b. What is the area of a square if its side measures $\sqrt{37}$ in?

12. **a.** Sketch a square with an area of 18 square centimeters.

b. What is its perimeter, to two decimal digits?

13. **a.** Sketch a square with a perimeter of 18 cm.

b. What is its area, to two decimal digits?

14. Place the letter of each expression in the box below its value, solving the riddle.

R $\sqrt{4.41}$	**O** $\sqrt{1.2 \cdot 0.3}$	**E** $\sqrt{0.16}$	**H** $\sqrt{169}$
G $\sqrt{100}$	**T** $\sqrt{\frac{225}{9}}$	**Q** $\sqrt{6^2 - 1}$	**R** $\sqrt{10^2 - 6^2}$
S $\sqrt{\frac{25}{4}}$	**E** $\sqrt{\frac{64}{25}}$	**U** $\sqrt{900}$	**S** $\sqrt{\frac{49}{100}}$
V $\sqrt{144}$	**E** $\sqrt{1.21}$	**A** $\sqrt{100(20 + 5)}$	**I** $\sqrt{41 \cdot 41}$
O $\sqrt{43 + 51}$	**M** $\sqrt{0.0016}$	**S** $\sqrt{10{,}000}$	**T** $\sqrt{4(20 - 8)}$

Why do plants hate math?

Because it...

10	41	12	0.4	2.5		5	13	1.1	0.04		100	$\sqrt{35}$	30	50	2.1	1 3/5		8	$\sqrt{94}$	0.6	$\sqrt{48}$	0.7

Make number 19 on a broken calculator that only has these buttons:

Irrational Numbers

The square roots of perfect squares are whole numbers. However, most numbers, such as 2, 5, and 17, are not perfect squares. We can find a **decimal approximation** to these types of square roots, either with a calculator, or manually, by using the techniques of squaring and guess-and-check.

Example 1. Find the value of $\sqrt{19}$ to two decimal digits, without using a calculator's square root function.

First we find two consecutive perfect squares so that 19 is between them: **16** < 19 < **25**. From that we know that $4 < \sqrt{19} < 5$. Also, since 19 is closer to 16 than to 25, we expect $\sqrt{19}$ to be closer to 4 than to 5.

So let's choose 4.3 and 4.4 as our initial guesses for the value of $\sqrt{19}$, square the guesses, and check how close to 19 we get.

Low Guess	$(LG)^2$	$(HG)^2$	High Guess
4.3	18.49	19.36	4.4

Note: $(LG)^2$ means low guess squared, and $(HG)^2$ means high guess squared.

From the table above, we can see that $\sqrt{19}$ is indeed between 4.3 and 4.4, and that it is probably closer to 4.4 than it is to 4.3 (because 19.36 is closer to 19 than 18.49 is). Let's try 4.36 and 4.37 next.

Low Guess	$(LG)^2$	$(HG)^2$	High Guess
4.3	18.49	19.36	4.4
4.36	19.0096	19.0969	4.37

Oops! $\sqrt{19}$ is not between 4.36 and 4.37. Both of those are too high. Let's try 4.35 and 4.36 next.

Low Guess	$(LG)^2$	$(HG)^2$	High Guess
4.3	18.49	19.36	4.4
4.35	18.9225	19.0096	4.36

Now we know that $\sqrt{19}$ is between 4.35 and 4.36 and closer to 4.36 than it is to 4.35 (because 19.0096 is much closer to 19 than 18.9225 is). This means that **to two decimal digits, $\sqrt{19}$ = 4.36.**

1. Continue refining the decimal approximation to $\sqrt{19}$, to three decimal digits. You may use a calculator to multiply, but do not use its square root function.

Low Guess	$(LG)^2$	$(HG)^2$	High Guess
4.35	18.9225	19.0096	4.36

2. Use only multiplication (squaring) to guess and check the values of the following square roots to two decimal digits. You may use a calculator, but not the square root function of the calculator.

a. $\sqrt{7}$

Low Guess	$(LG)^2$	$(HG)^2$	High Guess

b. $\sqrt{51}$

Low Guess	$(LG)^2$	$(HG)^2$	High Guess

c. $\sqrt{99}$

Low Guess	$(LG)^2$	$(HG)^2$	High Guess

3. Sarah claims that $\sqrt{11}$ equals exactly 3.317. Is she correct? Explain.

4. Prove that $\sqrt{2}$ cannot equal the fraction $\frac{71}{50}$.

Recall that **a rational number** is a number that **can be written as a ratio** (or as a fraction) **of two integers**, with a nonzero denominator. For example, 14/13 is a rational number, and so are −6 (= −6/1) and 7.89 (= 789/100).

An **irrational number** is just the opposite: one that ***cannot*** **be written as a ratio of two integers**.

Each rational number, being a fraction, can also be written as a decimal. The decimal form of a rational number either ends, or is unending with a repeating pattern in its digits.

The decimal expansion of an irrational number is unending and has no repeating pattern in the decimal digits.

Examples.

- The rational number 2/5 as a decimal is 0.4 — a decimal that ends.
- The fraction 1/3 = 0.3333... with 3 repeating has an unending decimal expansion. We can also write $1/3 = 0.\overline{3}$ where the line over the digit 3 means that that digit repeats indefinitely.
- $1.4\overline{08}$ means 1.408080808... with "08" repeating. Since the decimal expansion repeats, this is a rational number, and thus can be written as a fraction: it is actually 1394/990.
- The number π is the most famous irrational number. Its decimal expansion starts out like this: 3.14159265358979323846264338327950..., and never has any repeating pattern.
- Another famous irrational number is the Golden Ratio: $(1 + \sqrt{5})/2$, with an approximate value of 1.618.
- If a whole number is not a perfect square, its square root is an irrational number. So, $\sqrt{28}$, $\sqrt{97}$, and $\sqrt{45}$ all are irrational numbers. And there are many others!

Example 2. Determine whether the numbers $\frac{\sqrt{5}}{2}$ and 0.989898 are rational or irrational.

The first, $\sqrt{5}/2$, is irrational. This is because $\sqrt{5}$ is irrational, and if you divide an irrational number by a rational number, you cannot get a rational number. (See the answer key for Puzzle Corner for proof.)

The same is true for other operations: if x is irrational and r is rational, then all of these are irrational: $x + r$, $x - r$, xr, and x/r. It follows that, for example, 7π is irrational.

The second, 0.989898, is a decimal that ends, so it is rational. (As a fraction, it is 989,898/1,000,000.)

5. Determine whether the following numbers are rational or irrational, and explain why.

a. 0.928

b. $\sqrt{128}$

c. $0.55\overline{812}$

d. 6.050606

e. $\pi/2$

f. $\sqrt{100}$

g. $0.20\overline{8}$

h. $5\sqrt{3}$

i. $\frac{\sqrt{15}}{3}$

j. $\sqrt{\frac{25}{4}}$

6. This is Henry's work. Some of it is in error — but he can learn! Correct the ones are wrong, and explain why.

a. 1.272727 is rational because its decimal expansion repeats.

b. $3\sqrt{49}$ is irrational because it has a square root.

c. $\frac{\pi}{3}$ is rational because it is a fraction.

d. $\frac{15}{4}$ is rational because it is a fraction.

7. Place the numbers in the correct places in the diagram of *real numbers* = the set of both rational and irrational numbers. Note: the set of whole numbers is {0, 1, 2, 3, 4, 5, ...}.

$6, -76, \frac{22}{7}, \sqrt{81}, \sqrt{8}, 4.05, \frac{\pi}{7}, \sqrt{2}, 0.2\overline{7}, \frac{\sqrt{21}}{8}, 0, -\frac{1}{8}, \sqrt{7}+1, -\frac{45}{9}, 3\sqrt{3}, \frac{5}{\sqrt{100}}, \sqrt{900}$

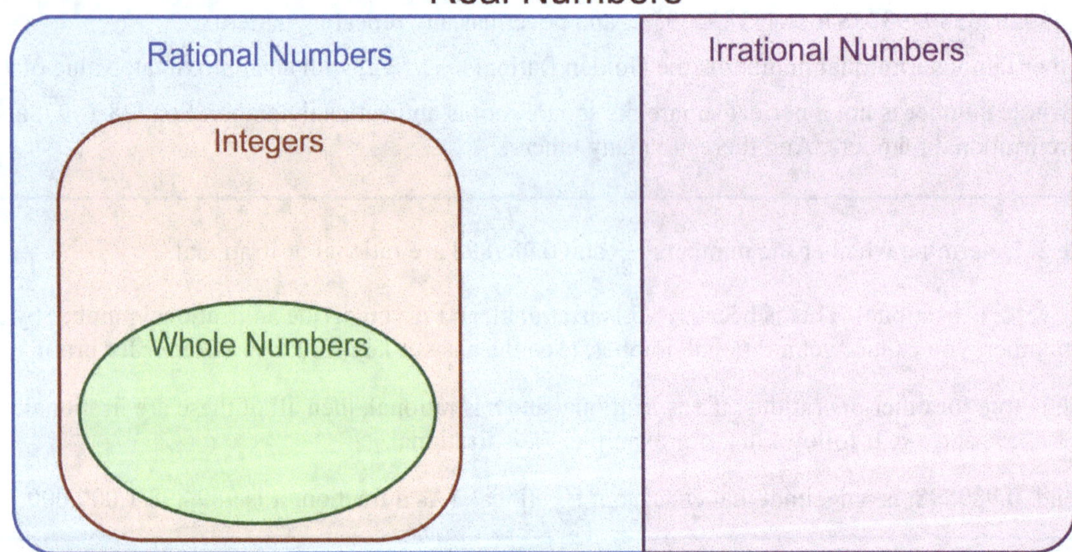

a. Read the proof that if x is irrational and r is rational, then $x + r$ is irrational.

Proof. Suppose the contrary, that $x + r = s$ where s is rational. Then, $x = s - r$. Since s and r can be written as fractions, their difference, $s - r$, is also a fraction, or a rational number.

This would mean x is rational, which is a contradiction. Therefore our original supposition cannot be correct. Thus, $x + r$ must be irrational.

b. Prove that an irrational number divided by a rational number is irrational.

Cube Roots and Approximations of Irrational Numbers

Similarly to the square root, we can take a **cube root** of a number.

Recall that the **cube of a number** is that number multiplied by itself three times. For example, two cubed = $2^3 = 2 \cdot 2 \cdot 2 = 8$. This gives us the volume of a cube with edges 2 units long.

The cube root of 8 is 2. We write it as $\sqrt[3]{8} = 2$. Notice the little "3" that is added to the radical sign to signify a cube root.

Example 1. Since $(-3)(-3)(-3) = -27$, then $\sqrt[3]{-27} = -3$.

Like square roots, most cube roots are irrational numbers. When it comes to integers, only the cube roots of perfect cubes are rational; the rest are irrational.

1. Find the cube roots without a calculator.

a. $\sqrt[3]{27}$	**b.** $\sqrt[3]{125}$	**c.** $\sqrt[3]{64}$	**d.** $\sqrt[3]{1{,}000}$
e. $\sqrt[3]{1}$	**f.** $\sqrt[3]{216}$	**g.** $\sqrt[3]{27{,}000}$	**h.** $\sqrt[3]{-8}$
i. $\sqrt[3]{-1}$	**j.** $\sqrt[3]{-125}$	**k.** $\sqrt[3]{0}$	**l.** $\sqrt[3]{-8{,}000}$

2. **a.** The volume of a cube is 216 cm^3. How long is its edge?

b. What is $(\sqrt[3]{4})^3$?

c. If the edge of a cube measures 50 cm, find its volume.

d. If the volume of a cube is 729 in^3, find its surface area.

3. (optional) Find the cube roots of these fractions and decimals, without a calculator.

a. $\sqrt[3]{0.008}$	**b.** $\sqrt[3]{0.125}$	**c.** $\sqrt[3]{-0.027}$
d. $\sqrt[3]{\frac{8}{125}}$	**e.** $\sqrt[3]{\frac{64}{27}}$	**f.** $\sqrt[3]{-\frac{1}{8}}$

Example 2. We can know that $\sqrt{98}$ lies between 9 and 10, because $9 = \sqrt{81} < \sqrt{98} < \sqrt{100} = 10$.

We can even tell it is much closer to 10 than to 9, since 98 is much closer to 100 than to 81.

From that, we can estimate that $2\sqrt{98}$ is slightly less than 20, and that $\sqrt{98} + 4$ is slightly less than 24.

Example 3. The opposite of $\sqrt{2}$ is $-\sqrt{2}$. Since $\sqrt{2}$ is approximately 1.41, then $-\sqrt{2} \approx -1.41$.

4. Find between which two whole numbers the root lies. Notice some of them are cube roots.

a. _____ $< \sqrt{31} <$ _____	**b.** _____ $< \sqrt{65} <$ _____	**c.** _____ $< \sqrt{87} <$ _____
d. _____ $< -\sqrt{5} <$ _____	**e.** _____ $< -\sqrt{44} <$ _____	**f.** _____ $< -\sqrt{50} <$ _____
g. _____ $< \sqrt[3]{7} <$ _____	**h.** _____ $< \sqrt[3]{37} <$ _____	**i.** _____ $< \sqrt[3]{101} <$ _____

5. Plot the following numbers *approximately* on the number line. Do not use a calculator, but think about between which two integers the root lies, and whether it is close to one of those integers.

$\sqrt{15}$ $\quad \sqrt{47}/2$ $\quad \sqrt[3]{9}$ $\quad -\sqrt[3]{27}$ $\quad -\sqrt{10}$ $\quad \sqrt{66}/2$ $\quad \pi$ $\quad \sqrt{18} + 1$

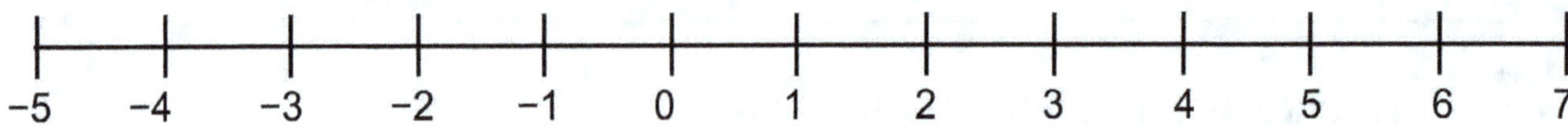

6. Compare, writing >, <, or = between the numbers. Think between which two whole numbers the root lies, using mental math.

a. 5 ☐ $\sqrt{27}$	**b.** $\sqrt{48}$ ☐ 7	**c.** $\sqrt{18}$ ☐ 4	**d.** $\sqrt[3]{9}$ ☐ 2
e. 2 ☐ $\sqrt{2} + 1$	**f.** $\sqrt{32} + 1$ ☐ 6	**g.** $\sqrt{43} + 5$ ☐ 10	**h.** $\sqrt{88} - 3$ ☐ 7

7. **a.** Between which two whole numbers does $\sqrt{30}$ lie? And $\sqrt{60}$?

b. Use your answers to (a) to determine whether $2\sqrt{30}$ is equal to $\sqrt{2 \cdot 30}$.

8. Is $\frac{\sqrt{50}}{2}$ equal to $\sqrt{\frac{50}{2}}$? Explain your reasoning.

9. Use the decimal approximations of common irrational numbers on the right to estimate the value of the expressions below, to one decimal digit. Use mental math and paper-and-pencil calculations, not a calculator.

$\pi \approx 3.14$

$\sqrt{2} \approx 1.41$

$\sqrt{5} \approx 2.24$

a. $5\sqrt{2}$ **b.** π^2 **c.** $\sqrt{5} - \sqrt{2}$ **d.** $2\sqrt{5} - 5\sqrt{2}$

10. **a.** Find an approximation to $\sqrt{11}$ to one decimal digit, without using the square root function of a calculator.

b. Use the approximation you found to estimate the values of $\sqrt{11} - \sqrt{2}$ and $3\sqrt{11}$.

11. Sarah has used the method of squaring her guesses to find out that $\sqrt{45}$ is between 6.7 and 6.8. How can she continue from this point to get a better approximation? Do it for her, to two decimal digits.

Use these exercises for additional practice.

12. Order the numbers from smallest to greatest. Estimate the value of the roots, thinking between which two whole numbers each square root lies, using mental math.

$\sqrt{5} - 1$ $\quad \sqrt[3]{1}$ $\quad \sqrt{19}/2$ $\quad \sqrt[3]{100}$ $\quad \sqrt[3]{8}$ $\quad \sqrt{13}$ $\quad \sqrt{9}$ $\quad 2\pi$ $\quad \sqrt{22} + 1$

13. Plot the following numbers *approximately* on the number line. Do not use a calculator, but think about between which two integers the root lies, and whether it is close to one of those integers.

a. $-2\sqrt{2}$ **b.** $\sqrt{80}/3$ **c.** $\sqrt{27} - 1$ **d.** $-\sqrt{5} + 7$

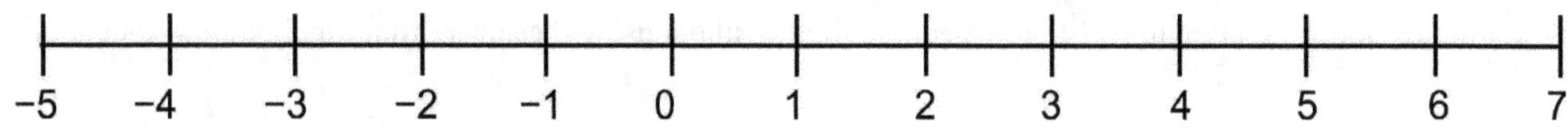

14. Plot the following numbers *approximately* on the number line.

a. $-\sqrt{2} - 3$ **b.** $-\pi$ **c.** $-\sqrt[3]{9}$ **d.** $-\sqrt{36} + 9$ **e.** $-\sqrt{26}/2$

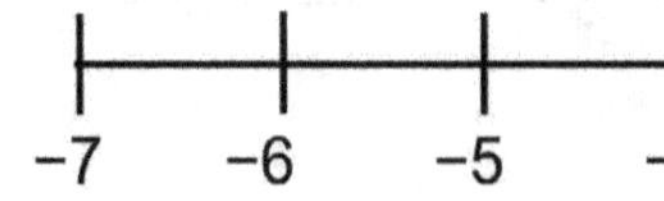
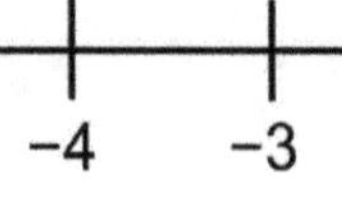
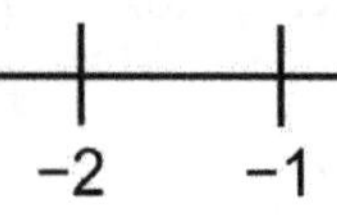
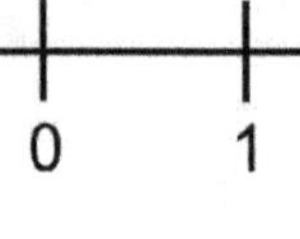
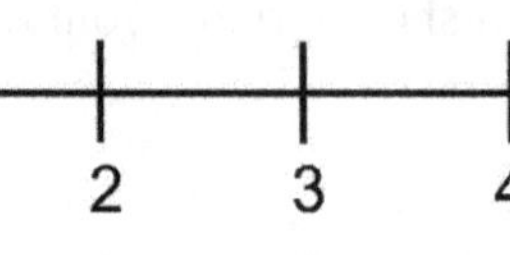

Fractions to Decimals

(This lesson is review, and optional.)

Each fraction is a rational number (by definition!). Each fraction can be written as a decimal. It will either be a terminating decimal, or a non-terminating repeating decimal.

It is easy to rewrite a fraction as a decimal when the denominator is a power of ten. However, when it is not (which is most of the time), simply treat the fraction as a division and divide. You will get either a **terminating decimal** or a non-terminating **repeating decimal**. See the examples below.

1. The denominator is a power of ten or the fraction can be simplified so that it is. In this case, writing the fraction as a decimal is straightforward. Simply write out the numerator. Then add the decimal point based on the fact that the number of zeros in the power of ten tells you the number of decimal digits.

Examples 1. $\frac{7809}{100} = 78.09$ $\frac{1458}{1000} = 1.458$ $\frac{506}{100{,}000} = 0.00506$ $\frac{33}{30} = \frac{11}{10} = 1.1$

2. The denominator is a factor of a power of ten. Convert the fraction into one with a denominator that is a power of ten. Then do as in case (1) above.

Examples 2. $\frac{9}{20} = \frac{45}{100} = 0.45$ $\frac{2}{125} = \frac{16}{1000} = 0.016$ $\frac{9}{8} = \frac{1125}{1000} = 1.125$

3. Use division (long division or with a calculator). This method works in all cases, even if the denominator happens to be a power of ten or a factor of a power of ten.

Example 3. Write $\frac{31}{40}$ as a decimal.

This division terminates (comes out even) after just three decimal digits.

We get $\frac{31}{40} = 0.775$. This is a **terminating decimal**.

(The fact the division was even means that the denominator 40 is a factor of some power of ten, and so we could have used method 2 from above. In this case, 1000 = 40 · 25.)

```
      0 0.7 7 5
40) 3 1.0 0 0 0
  - 2 8 0
      3 0 0
    - 2 8 0
        2 0 0
      - 2 0 0
            0
```

Example 4. Write $\frac{18}{11}$ as a decimal.

We write 18 as 18.0000 in the long division "corner" and divide by 11. Notice how the digits "63" in the quotient, and the remainders 40 and 70, start repeating.

So $\frac{18}{11} = 1.\overline{63}$.

The fraction 18/11 equals $1.\overline{63}$, which is a **repeating decimal**.

```
     0 1.6 3 6 3
11 ) 1 8.0 0 0 0
    -1 1
       7 0
     - 6 6
         4 0
       - 3 3
           7 0
         - 6 6
             4 0
           - 3 3
               7
```

1. Write the fractions as decimals.

a. $\frac{2}{100}$ =	**b.** $\frac{278}{10,000}$ =	**c.** $\frac{55,073}{1,000,000}$ =
d. $\frac{4508}{1000}$ =	**e.** $\frac{56,330}{100}$ =	**f.** $\frac{40,309}{10,000}$ =

2. Write the fractions as decimals.

a. $\frac{2}{5}$ =	**b.** $\frac{24}{25}$ =	**c.** $\frac{54}{200}$ =
d. $\frac{7}{4}$ =	**e.** $\frac{330}{250}$ =	**f.** $\frac{7}{125}$ =

3. Write as decimals. Use long division, and calculate each answer to at least six decimal places. If you find a repeating pattern, give the repeating part. If you don't, round your answer to five decimals.

a. $\frac{5}{9}$	**b.** $\frac{508}{27}$	**c.** $\frac{23}{61}$

Decimals to Fractions

When a decimal terminates, it is straightforward to write it as a fraction:

1. Copy all the digits (without a decimal point and the leading zeros) to be the numerator.
2. You will find the denominator by checking the number of decimal places: for two decimal places, the denominator is 100, for four decimal places, the denominator is 10,000, and for n decimal places, it is 10^n.

Example 1. $0.00507 = \frac{507}{100{,}000}$ $\quad 5.256 = \frac{5256}{1000}$ $\quad 0.0818 = \frac{818}{10{,}000}$

When a decimal does not terminate and has a repeating pattern in the decimal digits, it is a rational number, and we can write it as a fraction. The examples show you how.

Example 2. To write $0.2\overline{38}$ as a fraction, we multiply it by some power of ten in such a manner that when we subtract this multiple of $0.2\overline{38}$ and $0.2\overline{38}$, the repeating digits are eliminated in the subtraction.

Since there are *two* digits that repeat, we multiply $0.2\overline{38}$ by 10^2 or 100. The digits will repeat in $100 \cdot 0.2\overline{38}$ <u>in the same places</u> that they do in $0.2\overline{38}$. This means that when we subtract $100 \cdot 0.2\overline{38} - 0.2\overline{38}$, the repeating digits will be eliminated.

Let $x = 0.2\overline{38}$. We will calculate $100x$ and then subtract $100x$ and x. See the equations on the right.

Remember to line up the decimal points carefully, so that the digits that repeat will be in the same places.

We subtract the left sides of the equations, and also the right sides. This leaves $99x$ on the left side, and 23.6 on the right side. From this new equation $99x = 23.6$, we can solve x: it is 23.6/99.

$$
\begin{array}{rl}
100x = & 23.8383838\ldots \\
-\ x = & \ \ 0.2383838 38\ldots \\
\hline
99x = & 23.6
\end{array}
$$

$x = 23.6/99 = 236/990$

$= 118/495$

Lastly, we multiply both the numerator and the denominator of that expression by 10, to get 236/990. This can still be simplified to <u>**118/495**</u>.

Example 3. Write $0.41\overline{509}$ as a fraction.

Let $y = 0.41\overline{509}$. Since there are three repeating digits, we will multiply y by 10^3 or 1000, and then subtract $1000y$ and y.

See the process on the right. We get $y = \frac{41{,}468}{99{,}900}$.

$$
\begin{array}{rl}
1000y = & 415.09509509\ldots \\
-\ y = & \ \ \ 0.41509509509\ldots \\
\hline
999y = & 414.68
\end{array}
$$

$y = 414.68/999 = 41{,}468/99{,}900$

(Now, this *can* be simplified to 10,367/24,975 but since factorization takes time, you do not have to simplify the fractions in the exercises when the numerators and the denominators are large.)

1. Write as a fraction. Simplify if you can, but it is not required.

a. 93.82	**b.** 0.333	**c.** 2.05056
d. 61.098	**e.** 0.0000045	**f.** 4.932048

2. Are the decimals $0.\overline{2}$ and 0.222 the same? If not, what is their difference?

3. Write each repeating decimal as a fraction.

a. $0.\overline{4}$	**b.** $0.2\overline{1}$
c. $0.\overline{954}$	**d.** $2.5\overline{32}$

4. Match the fractions and the decimals.

0.666	$0.\overline{51}$	$0.\overline{6}$	$0.0\overline{6}$	0.51	0.051	0.066	$0.0\overline{51}$
$\frac{51}{100}$	$\frac{2}{3}$	$\frac{66}{1000}$	$\frac{66}{990}$	$\frac{51}{99}$	$\frac{51}{990}$	$\frac{666}{1000}$	$\frac{51}{1000}$

5. Erica and Eric were comparing their methods and results for writing the decimal $2.1\overline{7}$ as a fraction.

Erica said, "Since there is one repeating digit, I multiplied it by 10, to get $21.\overline{7}$, and then subtracted $10x - x$ to get $9x = 19.6$. So, I got $x = 19.6/9 = 196/90 = 98/45$."

Eric said, "I multiplied it by 100, and got $100x - x = 217.\overline{7} - 2.1\overline{7} = 215.6$. From that, $x = 215.6/99 = 2156/990$."

Who is correct?

6. Janet's calculator gave her the result that 305/55494 = 0.00549608966735142537932028687786.
Janet said that since there is no repeating pattern here, that this is an irrational number.
Is she correct?

7. Write each repeating decimal as a fraction.

a. $0.\overline{256}$	**b.** $3.05\overline{94}$
c. $0.03\overline{2199}$	**d.** $1.\overline{36309}$

Square and Cube Roots as Solutions to Equations

Example 1. Solve $x^2 = 81$.

We can use mental math: one obvious solution is $x = 9$. However, there is also another solution! It is not only true that $9^2 = 81$, but $(-9)^2 = 81$ also, so $x = -9$ is a second solution to this equation.

Example 2. Solve $x^2 = 48$.

This time, we cannot solve the equation with mental math, but we will *take a square root of both sides of the equation.* This will undo the squaring, and isolate x, because taking a square root and squaring are opposite operations.

$x^2 = 48$	$\sqrt{\ }$	The radicand symbol signifies taking a square root of both sides of the equation.
$x = \sqrt{48} \approx 6.93$		Since taking a square root undoes the squaring, x is now left alone on the left side. Notice that there are two solutions: the square root of 48 and the negative square root of 48.
or $x = -\sqrt{48} \approx -6.93$		

Notice that $-\sqrt{48}$ doesn't mean that we take a square root of a negative number. Instead, $-\sqrt{48}$ means we *first* take the square root of 48 (a positive number) and then take the opposite of that result.

1. Solve. Remember, there will be two solutions: one positive and one negative. When the two answers aren't integers, give them as square roots and also as decimals rounded to two decimal digits. 

a. $x^2 = 25$	**b.** $y^2 = 3{,}600$
c. $x^2 = 500$	**d.** $z^2 = 11$
e. $w^2 = 287$	**f.** $q^2 = 1{,}000{,}000$

The situation is similar with equations where the variable is cubed.

Example 3. Solve $x^3 = 125$.

Since 125 is a perfect cube, the solution is easy to find with mental math.

$$\begin{aligned} x^3 &= 125 \quad \Big| \sqrt[3]{\ } \\ x &= \sqrt[3]{125} = 5 \end{aligned}$$

With cube roots, **there is no other solution.** For example, in this case, $(-5)^3$ does not equal 125.

Example 4. Solve $x^3 = 35$.

We will take the cube root of both sides of the equation. This will undo the cubing and isolate x.

$$\begin{aligned} x^3 &= 35 \quad \Big| \sqrt[3]{\ } \\ x &= \sqrt[3]{35} \approx 3.27 \end{aligned}$$

Since 35 is not a perfect cube, $\sqrt[3]{35}$ is an irrational number. Depending on context, we might give the answer in root form, or with a decimal approximation.

Hint: If your calculator does not have a button for the cube root, you can instead use the button for exponentiation, with the exponent 1/3. For example, on my computer calculator, I enter $\sqrt[3]{23}$ this way:

2. Solve. If the root is not a whole number, give it rounded to two decimal digits.

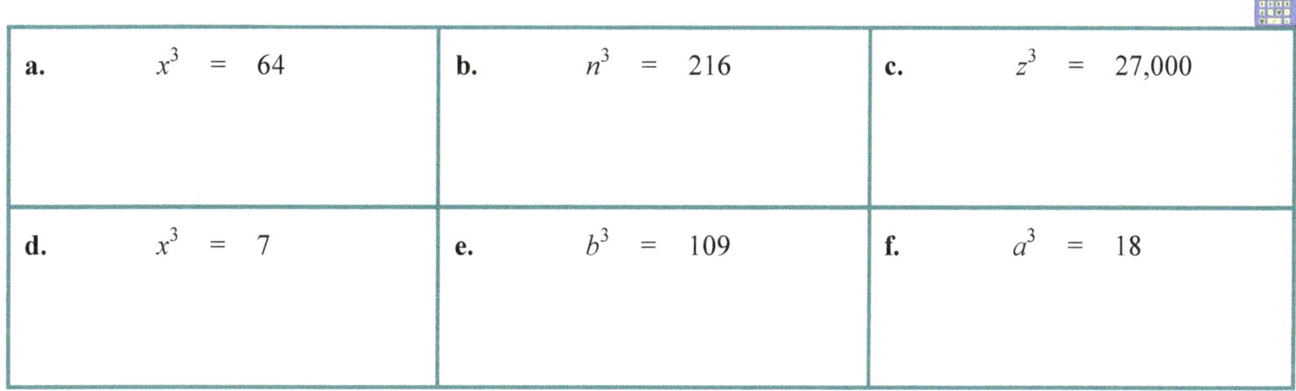

a. $x^3 = 64$	**b.** $n^3 = 216$	**c.** $z^3 = 27{,}000$
d. $x^3 = 7$	**e.** $b^3 = 109$	**f.** $a^3 = 18$

3. Below, the variable V signifies the volume of a cube. Find the edge of the cube (s) when the volume is given. Give the answer to the same amount of significant digits as the given volume.

a. $V = 510 \text{ m}^3$ $s =$	**b.** $V = 24{,}500 \text{ cm}^3$	**c.** $V = 5.83 \text{ ft}^3$

4. Find the surface area of a cube with a volume of 14.2 m^3.

Example 5. $x^2 + 78 = 129$ — We want to isolate the term x^2, so we first subtract 78 from both sides.

$x^2 = 51$ — Now we take a square root of both sides.

$x = \sqrt{51}$ or $x = -\sqrt{51}$ — There are two solutions, as usual.

If this is strictly a math problem and does not involve quantities with units, the answer can be left in the root form, which is exact. Otherwise, you should find its decimal approximation.

Here are the checks. Usually, it is enough to check only the positive root ($x = \sqrt{51}$), as the check for the negative root ($x = -\sqrt{51}$) is practically identical.

$$(\sqrt{51})^2 + 78 \stackrel{?}{=} 129$$
$$51 + 78 \stackrel{?}{=} 129$$
$$129 = 129 \checkmark$$

$$(-\sqrt{51})^2 + 78 \stackrel{?}{=} 129$$
$$51 + 78 \stackrel{?}{=} 129$$
$$129 = 129 \checkmark$$

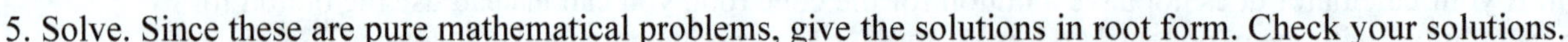

5. Solve. Since these are pure mathematical problems, give the solutions in root form. Check your solutions.

a. $a^2 - 8 = 37$	**b.** $y^2 + 100 = 1{,}000$
c. $b^2 + 1.5 = 6.4$	**d.** $x^2 - 26 = 709$

6. Solve. Give the solutions in exact form. Check your solutions.

a. $x^3 - 5 = 59$	**b.** $x^3 + 78 = 437$

More Equations that Involve Roots

Example 1.

$3x^2 = 40$	$\div 3$	Again, we want to isolate the term x^2, so we first divide both sides by 3.
$x^2 = 40/3$	$\sqrt{\ }$	This symbol signifies taking a square root of both sides of the equation.
$x = \sqrt{40/3}$ or $x = -\sqrt{40/3}$		There are two solutions, as usual.
$x \approx 3.651$ or $x \approx -3.651$		These are the decimal approximations.

Here is a check using the rounded positive root:

$$3 \cdot 3.651^2 \stackrel{?}{=} 40$$
$$3 \cdot 13.329801 \stackrel{?}{=} 40$$
$$39.989403 \approx 40 \checkmark$$

Here is a check using the exact positive root:

$$3 \cdot (\sqrt{40/3})^2 \stackrel{?}{=} 40$$
$$3 \cdot 40/3 \stackrel{?}{=} 40$$
$$40 = 40$$

1. Solve. Give the solutions both in exact format and as decimals rounded to three decimals.

a. $5x^2 = 125$	**b.** $8.2b^2 = 319$
c. $a^2 + 4.5 = 10.7$	**d.** $12b^2 = 36{,}000$

Example 2.

$x^2 + 7^2 = 12^2$		First simplify.
$x^2 + 49 = 144$		Now it looks more familiar. Subtract 49.
$x^2 = 95$		Now we take a square root of both sides.
$x = \sqrt{95}$ or $x = -\sqrt{95}$		There are two solutions, as usual.

Check:

$(\sqrt{95})^2 + 7^2$  12^2

$95 + 49 \stackrel{?}{=} 144$

$144 = 144$ 

2. Solve. Give the solutions in exact form. Check your solutions.

a. $a^2 + 3^2 = 7^2$	**b.** $43^2 + x^2 = 51^2$

3. Solve. Give the solutions rounded to two decimals. Check your solutions.

a. $s^2 = 2.1^2 + 5.4^2$	**b.** $21^2 - w^2 = 15^2$
c. $121 - x^2 = 56$	**d.** $a^2 - 4.5^2 = 5.78$

4. Solve equations involving cube roots, also. Give each solution rounded to three decimals.

a. $56 + s^3 = 542$	**b.** $5x^3 = 180$	**c.** $254 - z^3 = 46$

5. *(Challenge.)* These equations involve cube roots of negative numbers. Give each solution in exact form.

a. $x^3 = -27$	**b.** $w^3 = -343$	**c.** $4t^3 = -4$
d. $5x^3 + 3 = -27$	**e.** $-10r^3 = 10{,}000$	**f.** $54 - x^3 = -1$

6. Here are some more practice problems. Round the answers to three decimals.

a. $45 - x^2 = 20$	**b.** $112^2 + s^2 = 18{,}200$	**d.** $6{,}650 - y^2 = 70^2$

Puzzle Corner

Solve $x^2 - x = 0$.

The Pythagorean Theorem

You will now learn a very famous mathematical result, the Pythagorean Theorem, which has to do with the lengths of the sides in a right triangle. First, we need to study some terminology.

In a right triangle, the two sides that are perpendicular to each other are called **legs**. The third side, which is always the longest, is called the **hypotenuse**.

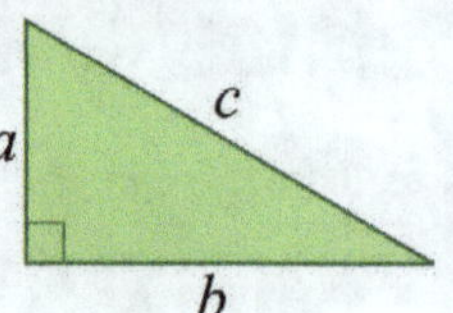

In the image on the right, the sides *a* and *b* are the legs, and *c* is the hypotenuse.

Note: We don't use the terms "leg" and "hypotenuse" to refer to the sides of an acute or obtuse triangle — this terminology is restricted to *right* triangles.

The Pythagorean Theorem states that **the sum of the squares of the legs equals the square of the hypotenuse.**

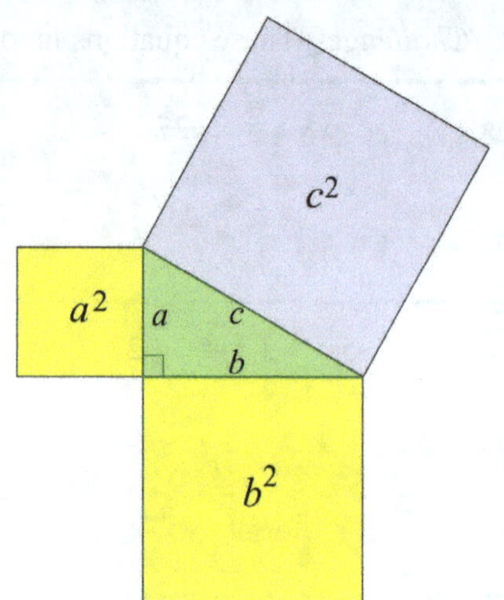

In symbols it looks much simpler:

$$a^2 + b^2 = c^2$$

The picture shows squares drawn on the legs and on the hypotenuse of a right triangle. Verify visually that the total area of the two yellow squares drawn on the legs looks about equal to the area of the blue square on the hypotenuse.

We will prove this theorem in another lesson.
For now, let's get familiar with it and learn how to use it.

1. Below you see the famous 3-4-5 triangle: its sides measure 3, 4, and 5 units and it is a right triangle. Check that the Pythagorean Theorem holds for it by filling in the numbers on the right.

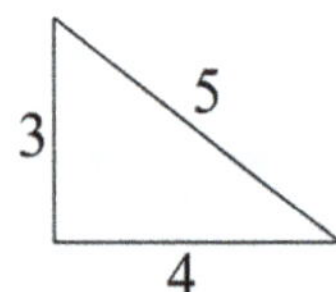

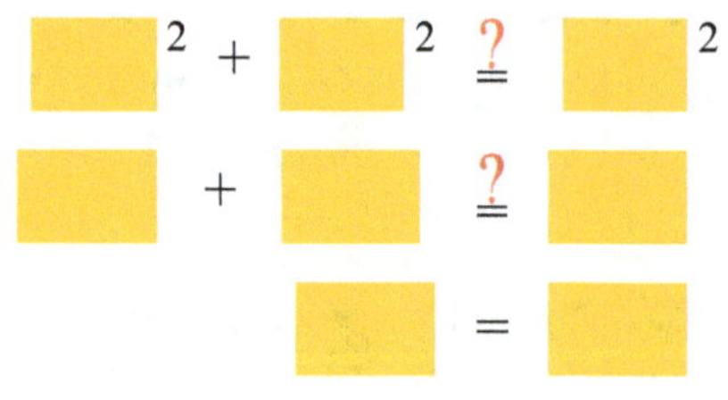

2. If we double the lengths from the 3-4-5 triangle, we get the 6-8-10 triangle. Use a ruler and a protractor to draw line segments that are 6 cm and 8 cm long and are perpendicular to each other.

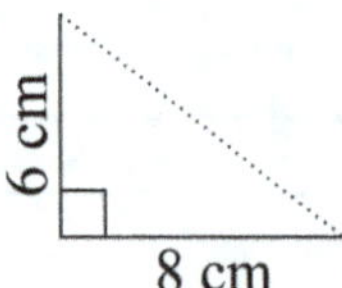

Then draw in the third side to complete the triangle. Does the hypotenuse of your triangle measure 10 cm? Do the lengths 6, 8, and 10 fulfill the Pythagorean Theorem?

Example 3. The two legs of a right triangle measure 7 and 10 units. How long is the hypotenuse?

Let x be the length of the unknown side, which is the hypotenuse. From the Pythagorean Theorem, we get:

$$7^2 + 10^2 = x^2$$
$$49 + 100 = x^2$$
$$x^2 = 149$$
$$x = \sqrt{149} \quad \sout{\text{or } x = -\sqrt{149}}$$

We ignore the negative root as the length of a side cannot be negative!

The answer is left in the root form since this is not a real-life problem.

3. Solve for the unknown side of each right triangle. Leave your answer in root form if the radicand (number under the radical) is not a perfect square.

a. 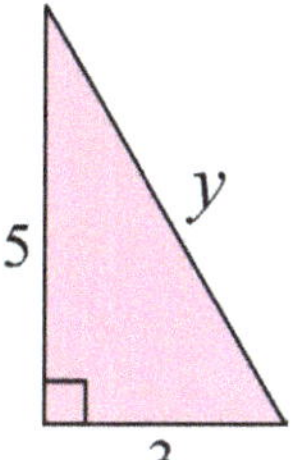	**b.** 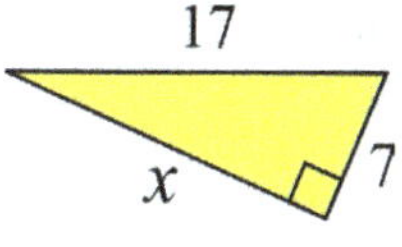
c. 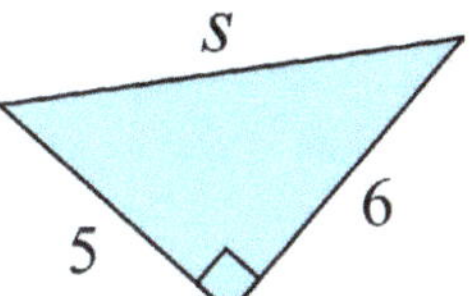	**d.** 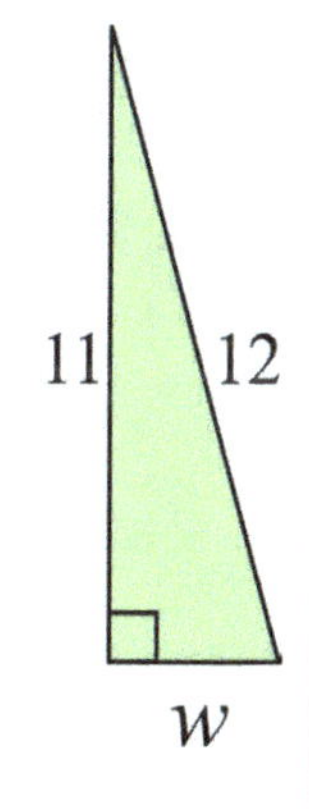

4. The sides of a square measure 6 units. How long is the diagonal of the square? Give your answer to two decimal digits.

Example 4. The hypotenuse of a right triangle is $\sqrt{27}$ and its one leg is 3. How long is the other leg?

Let y be the length of the unknown leg. According to the Pythagorean Theorem, we get:

$$y^2 + 3^2 = (\sqrt{27})^2$$
$$y^2 + 9 = 27$$
$$y^2 = 18$$
$$y = \sqrt{18}$$

Again, we do not include the negative root since the length of a side cannot be negative.

Being a purely mathematical problem, the answer is left in the root form.

5. Solve for the unknown side of each right triangle. Leave your answer in root form if the radicand (number under the radical) is not a perfect square.

a.

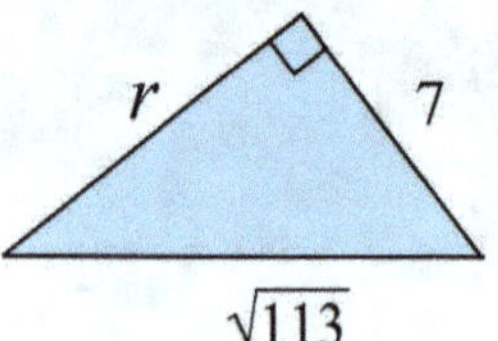

b.

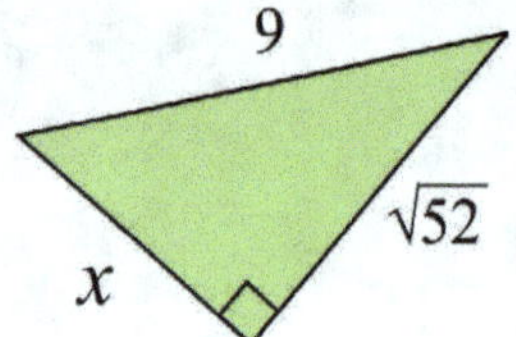

c.

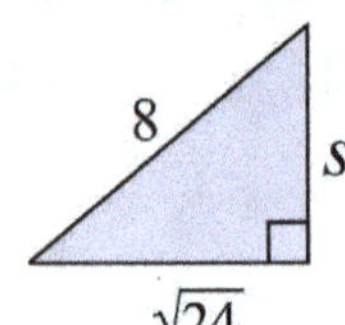

d.

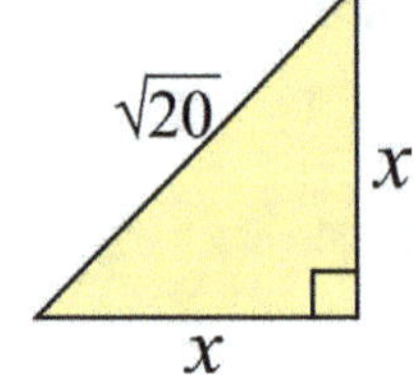

6. **a.** The two legs of a right triangle are $\sqrt{7}$ and $\sqrt{8}$. How long is the hypotenuse?

b. The hypotenuse of a right triangle is $\sqrt{67}$ and one leg is $\sqrt{41}$. How long is the other leg?

Example 5. Find the unknown leg of this right triangle.

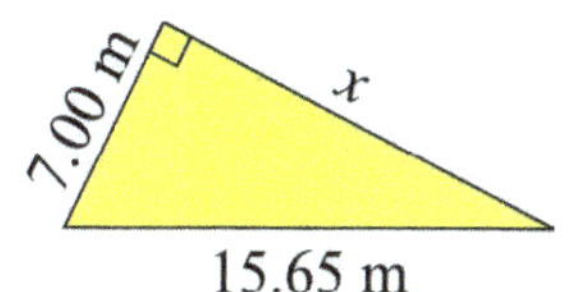

This time we know the hypotenuse and one of the legs.
The Pythagorean Theorem gives us:

$$7.00^2 + x^2 = 15.65^2$$

$$49 + x^2 = 244.9225$$

$$x^2 = 195.9225$$

We keep all the decimals for the intermediate results.

$$x = \sqrt{195.9225} \text{ or } \cancel{x = -\sqrt{195.9225}}$$

Again, we ignore the negative root.

$$x \approx 14.00 \text{ m}$$

The answer is given to the hundredth of a meter, just like the lengths given in the problem.

The other leg measures about 14.00 m.

7. Solve for the unknown side. Round your answer to the same accuracy as the numbers in the problem.

a.

37.0 cm
w
42.1 cm

b.

1044 ft
t
1131 ft

c.

51 m
x
17 m

8. If the legs of a right triangle measure 12 ft 5 in and 7 ft 8 in, find the length of the hypotenuse to the nearest inch.

9. **a.** Measure each side of this triangle to the nearest millimeter.

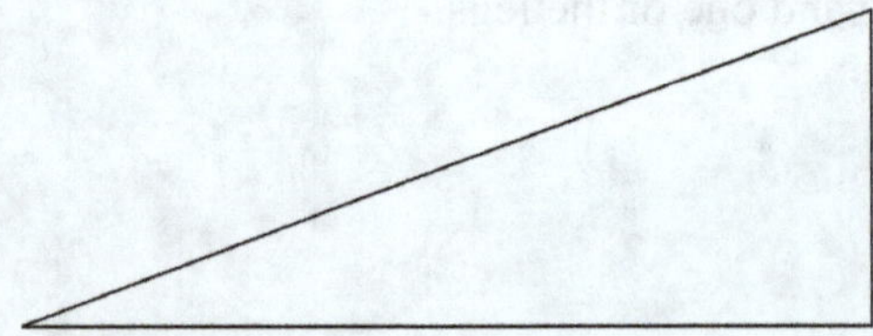

b. Verify that the sum of the areas of the squares on the legs is *very close* to the area of the square on the hypotenuse. I say "very close" because the process of measuring is always inexact, and therefore your calculations and results will probably not yield true equality, just something close.

10. How long is the diagonal of a laptop screen that is 9.0 inches high and 14.4 inches wide?

Note: when a laptop is advertised as having a "15-inch screen," it is the *diagonal* that is 15 inches, not the width or the height.

Puzzle Corner

A math teacher made the problem below for a test. Find what went wrong with it. Then fix the problem, so it can be used in the test, and solve it.

How long is the unknown side?

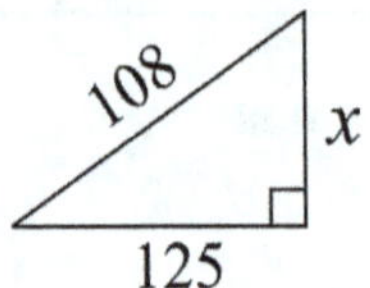

Applications of the Pythagorean Theorem 1

Example 1. An eight-foot ladder is placed against a wall so that the base of the ladder is 2 ft away from the wall. What height does the top of the ladder reach?

Since the ladder, the wall, and the ground form a right triangle, this problem is easily solved by using the Pythagorean Theorem. Let h be the unknown height. From the Pythagorean Theorem, we get:

$$\begin{aligned} 2^2 + h^2 &= 8^2 \\ 4 + h^2 &= 64 \\ h^2 &= 60 \\ h &= \sqrt{60} \\ h &\approx 7.75 \end{aligned}$$

Our answer, 7.75, is in feet. This means the ladder reaches to about 7 3/4 ft = 7 ft 9 in. high.

1. The area of a square is 100.0 m^2. How long is the diagonal of the square?

2. A park is in the shape of a rectangle and measures 48 m by 30 m. How much longer is it to walk from A to B around the park than to walk through the park along the diagonal path?

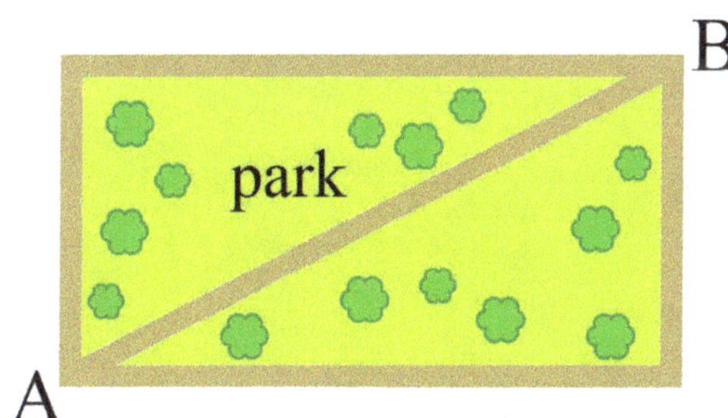

3. A clothesline is suspended between two apartment buildings. Calculate its length, assuming it is straight and doesn't sag any.

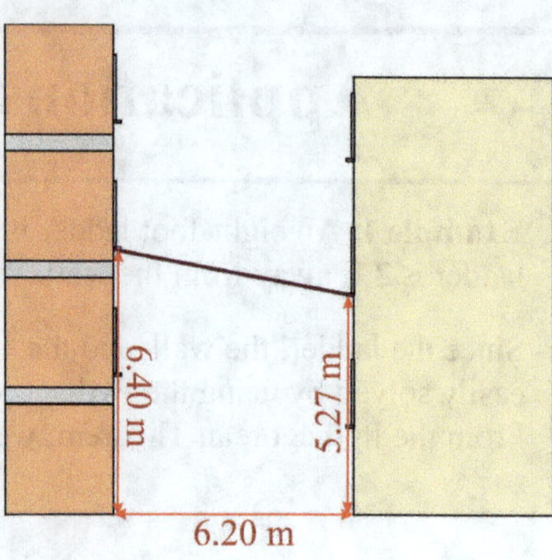

4. Find the perimeter of triangle ABC to the nearest tenth of a unit.

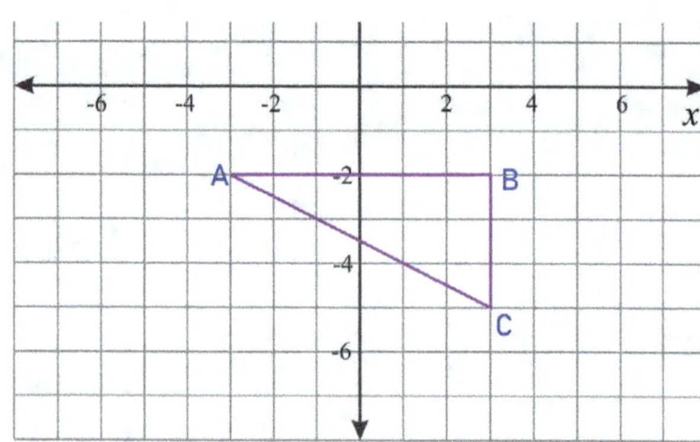

Example 2. Find the area of an isosceles triangle with sides 92 cm, 92 cm, and 40 cm.

Solution: To calculate the area of any triangle, we need to know its altitude. When we draw the altitude, we get a right triangle:

The next step is to apply the Pythagorean Theorem to solve for the altitude *h*, and after that calculate the actual area.

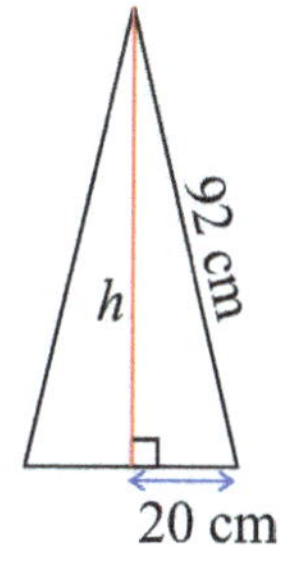

5. Calculate the area of the isosceles triangle in the example above to the nearest ten square centimeters.

6. Calculate the area of an equilateral triangle with 24-cm sides to the nearest square centimeter. Don't forget to draw a sketch.

A Proof of the Pythagorean Theorem and of Its Converse

There exist hundreds of different proofs for the Pythagorean Theorem. In this lesson, we will look at two geometric proofs.

Proof.

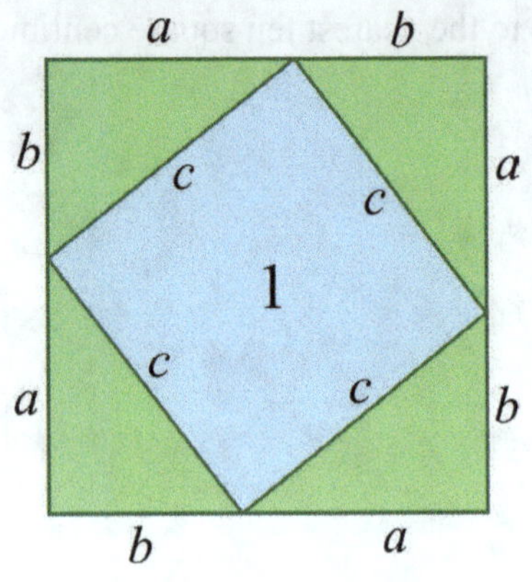

The figure above has four right triangles, each with sides a, b and c. The sides of the outside square are $a + b$. The triangles enclose a square with sides c units long.

Here the sides of the large square are still $a + b$, but the four right triangles have been rearranged so that two smaller squares are formed, one with side a and the other with side b.

Since the areas of both large squares are equal, and the areas of the four right triangles are equal, it follows that the remaining (blue) areas are also equal. In other words, the area of square 1, which is c^2, equals the area of square 2 (which is a^2) plus the area of square 3 (which is b^2). In symbols, $c^2 = a^2 + b^2$. ☺

1. Figure out how this proof of the Pythagorean Theorem works.

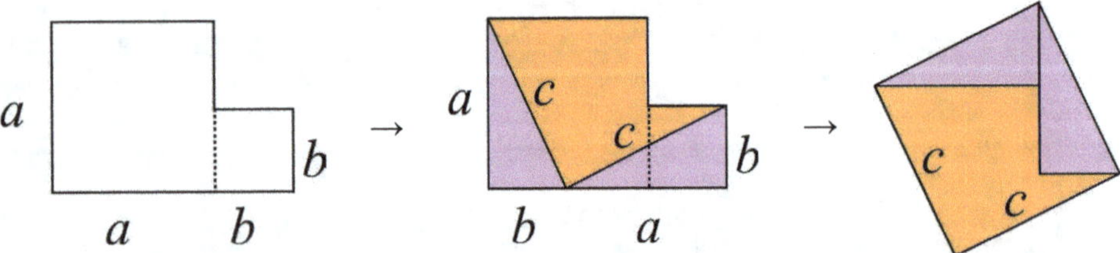

2. Study one of the proofs enough so you can explain it to someone else. Then do so.

Often in mathematics, **theorems** are of the form "If (statement 1), then (statement 2)."
The **converse** of such a theorem is, "If (statement 2), then (statement 1)."

For example, here is a true theorem that follows this "*if-then*" format:

"If the corresponding angles in two triangles are congruent, then the triangles are similar."

Its converse is: "If two triangles are similar, then their corresponding angles are congruent" , which is also true.

But sometimes, a mathematical statement is only true in one way, and its converse is not true. Here is an example of that.

Theorem: "If the sides of two rectangles are congruent, their areas are equal." (true)

Its converse: "If two rectangles have equal areas, their sides are congruent." (not true)

If we write the Pythagorean Theorem in this format of "If (statement 1), then (statement 2)", it becomes:

If a triangle with sides a, b, and c is a right triangle, then its sides fulfill the equation $a^2 + b^2 = c^2$.

Its converse is:

If the sides a, b, and c of a triangle fulfill the equation $a^2 + b^2 = c^2$, then the triangle is a right triangle.

3. Tell the converse of each statement. Then tell whether the original statement and its converse are true or not.

a. STATEMENT: If there are no clouds on the sky, then it is not raining. true / false

CONVERSE: true / false

b. STATEMENT: If you eat spoiled food, you will get gastrointestinal symptoms. true / false

CONVERSE: true / false

c. STATEMENT: If you just had your 7th birthday, then you'll turn 8 in a year. true / false

CONVERSE: true / false

d. STATEMENT: If $a + b = c$, then $c - b = a$. true / false

CONVERSE: true / false

e. STATEMENT: If you multiply two odd numbers, the product is also odd. true / false

CONVERSE: true / false

The **converse of the Pythagorean Theorem** states that if the relationship $c^2 = a^2 + b^2$ is true for the side lengths *a*, *b*, *c* of some triangle, then the triangle is a right triangle.

Proof. This is a proof by contradiction. This means that we start by assuming the statement we're trying to prove is *not* true, and then try to derive a contradiction. If we succeed, then it follows that the original statement must be true.

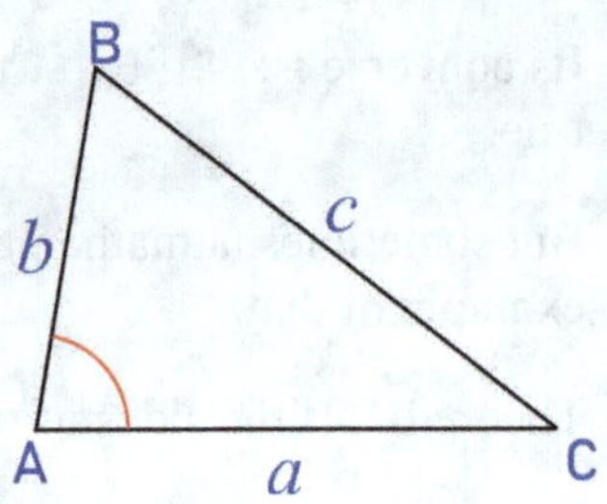

Let's say that the relationship $c^2 = a^2 + b^2$ is true for the side lengths *a*, *b*, *c* of triangle ABC, but the triangle is *not* a right triangle.

Let's further assume the angle at A is an acute angle. (The proof will work almost identically if it was obtuse.)

Now, we will draw a line segment of the length *a* so that the angle BAC' is a right angle, as in the image below:

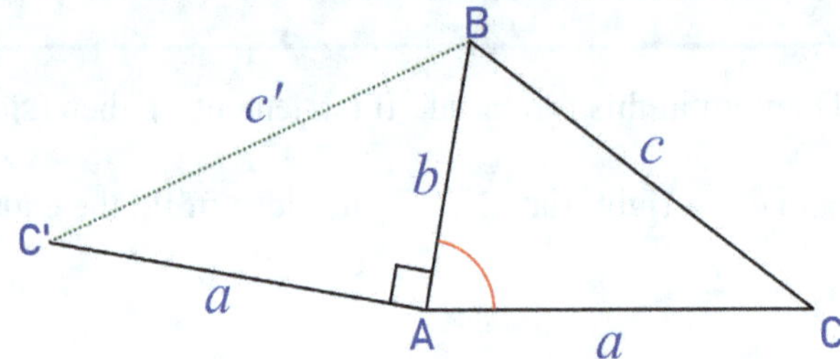

This forms the *right* triangle ABC', which shares the side $\overline{AB}$ with triangle ABC. The side lengths of this triangle are *a*, *b*, and *c*'.

Now, since triangle ABC' is a right triangle, the Pythagorean Theorem holds, and thus, $a^2 + b^2 = c'^2$. And our original assumption was that $a^2 + b^2 = c^2$.

From those two equations, it follows that $c^2 = c'^2$, and since *c* and *c*' cannot be negative, it follows that $c = c'$. So, the two triangles have the same lengths of sides: *a*, *b*, and *c*.

Yet, with three given lengths, you can only make *one* unique triangle, not two that would have different angles. This is a contradiction. Thus, our assumption that ABC is not a right triangle cannot be true.

Hence, ABC is a right triangle, and the converse of the Pythagorean Theorem is true.

4. Study the proof and make sure you understand it. Study it enough that you can tell it to someone else. Then tell it to someone else.

5. Is this corner a right angle?

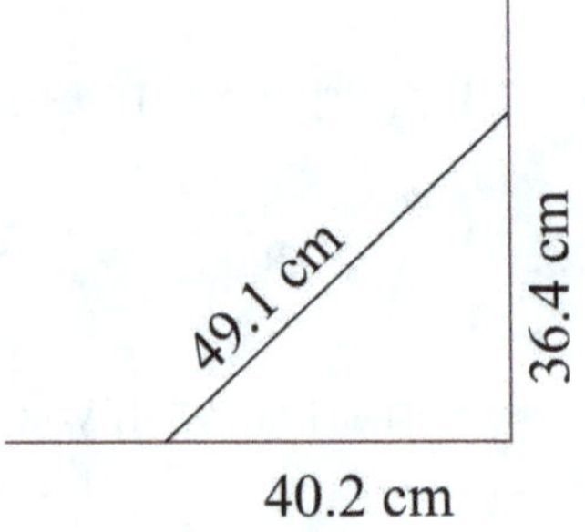

6. Construction workers have made a (hopefully) rectangular mold out of wood, and they are getting ready to pour cement into it. How could they make sure that the mold is indeed a rectangle and not a parallelogram? After all, in a parallelogram the opposite sides are equal, so simply measuring the opposite sides does not guarantee that a shape is a rectangle.

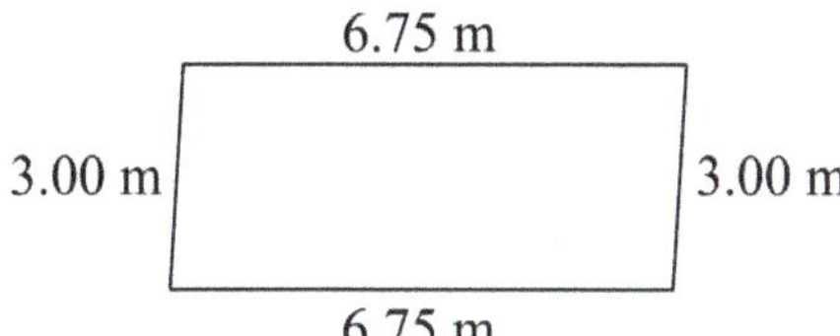

Example 1. (optional) This triangle is *not* a right triangle, so the Pythagorean Theorem does *not* hold:

$$2.55^2 + 3.31^2 \overset{?}{=} 3.58^2$$

$$6.5025 + 10.9561 \overset{?}{=} 12.8164$$

$$17.4586 > 12.8164$$

(In the image, "u²" signifies a square unit.)

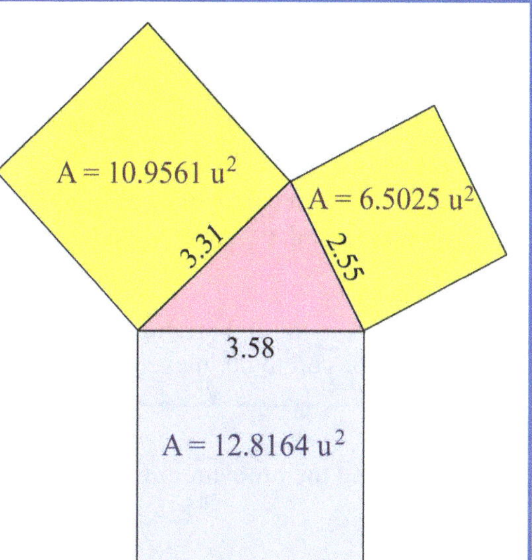

The sum of the areas of the squares drawn on the two shortest sides is *more* than the area of the square drawn on the longest side. As you can see, the triangle is acute.

If $a^2 + b^2 < c^2$ (where c is the longest side), then the triangle is obtuse.

7. (optional) For each set of lengths below, determine whether the lengths form an acute, right, or obtuse triangle. You can use the Pythagorean theorem, and/or construct the triangles using a compass and a ruler. To learn or to review how to do the latter, check this video — the fifth example in it explains how to draw a triangle with three given sides: https://www.mathmammoth.com/videos/geometry/draw_triangles

a. 6, 9, 13	**b.** 12, 13, 5	**c.** 4, 5, 7
d. 4, 5, 6	**e.** 13, 11, 10	**f.** 15, 20, 25

Applications of the Pythagorean Theorem 2

Example 1. Find the diagonal of the rectangular prism on the right.

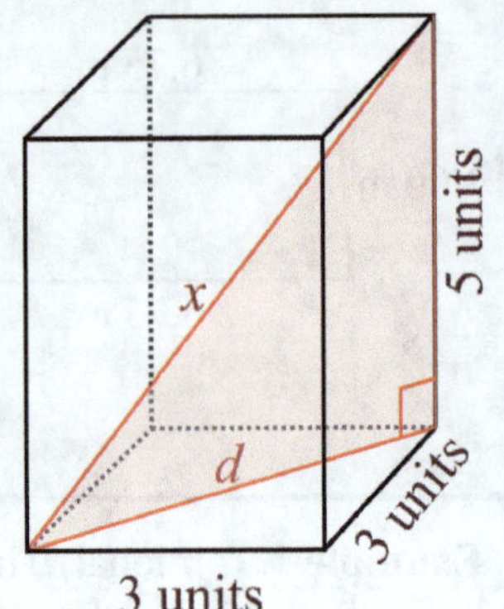

The diagonal in question is marked with x. To find its length, we will use the right triangle shaded in pink.

That right triangle, in its turn, has as one of its legs the diagonal (d) of the bottom square of the prism.

We will first solve for d using the Pythagorean Theorem:

$$3^2 + 3^2 = d^2$$
$$18 = d^2$$
$$d = \sqrt{18}$$

We don't need a decimal approximation for d, since this is only an intermediary result. (Since we continue the calculation, it is better to use the exact value, but if not, you would want to keep at least 4 decimals.)

Next, we look at the right triangle with sides 5, d or $\sqrt{18}$, and x, and use the Pythagorean Theorem. This is left for you to do in exercise 1.

1. Finish solving the problem in Example 1. Give your answer in root form.

2. Find the diagonal of a cube with 15-cm edges.
   Draw a sketch first.

3. The picture shows a right pyramid with a 12-inch square as the base.
   Find its surface area. (You will need to draw additional lines to the picture.)

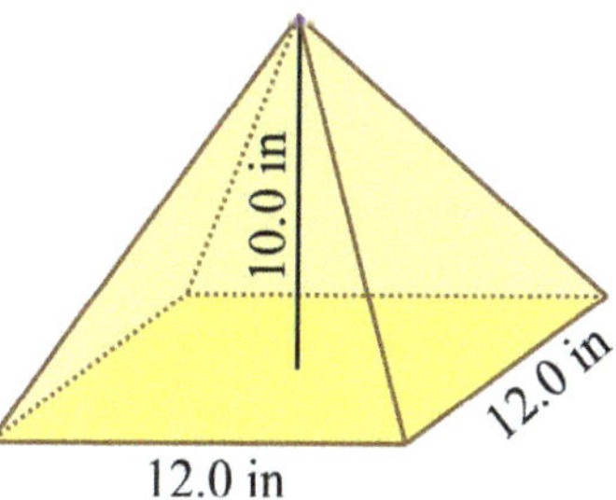

4. Calculate the length of the rafter in feet and inches, if...

   a. ...the run is 12 ft and the rise is 3 ft

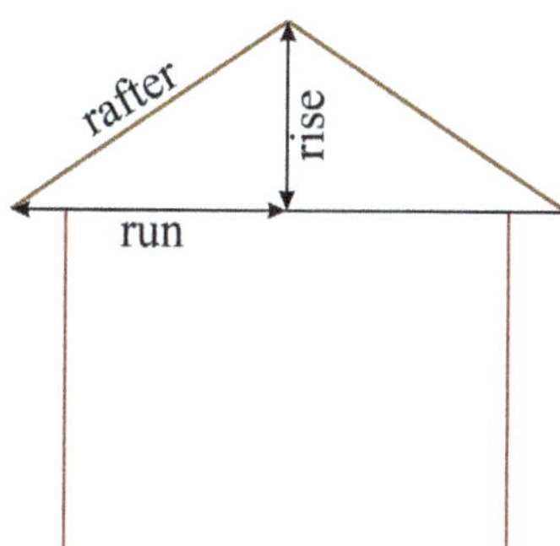

   b. ...the run is 12 ft and the rise is 5 ft 3 in.

5. Find the surface area of this roof to the nearest tenth of a square meter.

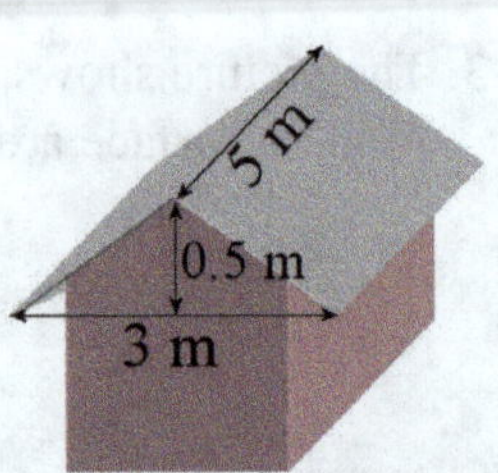

6. The roof of a little kiosk is in the shape of a square pyramid. Each bottom edge measures 3.5 m, and the other edges measure 3.2 m. Find the surface area of the roof to the nearest tenth of a square meter.

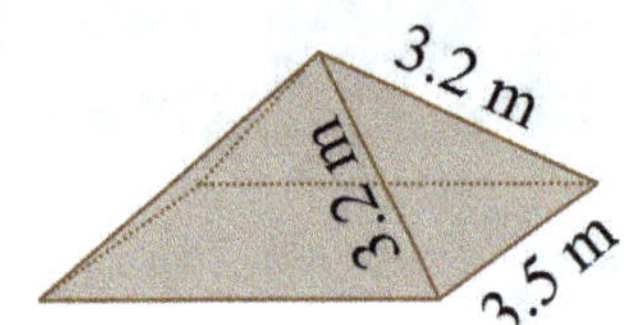

7. A creek runs through a piece of land in a straight line.

a. Find the length of the creek. Give your answer to the same accuracy as the dimensions in the picture.

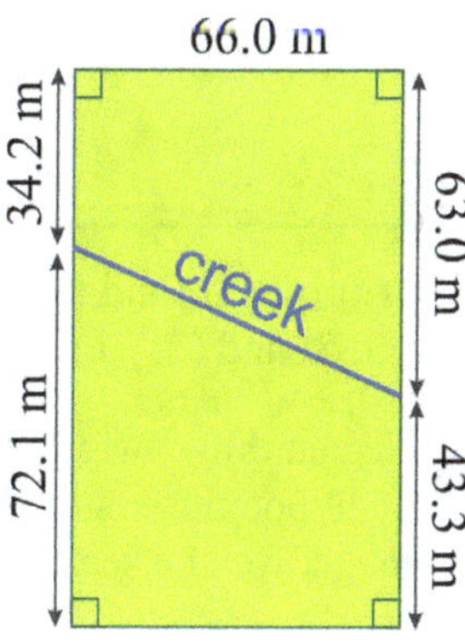

b. The creek splits the plot into two parts. Calculate the areas of the two parts.

Puzzle Corner The hypotenuse of a right triangle measures 12.0 m, and the one leg is twice the length of the other. Find the side lengths of the triangle.

Distance Between Points

Example 1. Find the distance between the points (−3, 5) and (2, −1).

We can draw a right triangle so that the length of the hypotenuse is the desired distance. From the image, we can see the legs are 5 and 6 units long.

Now it is easy to use the Pythagorean Theorem to find the distance:

$$5^2 + 6^2 = x^2$$
$$25 + 36 = x^2$$
$$61 = x^2$$
$$x = \sqrt{61} \approx 7.8 \text{ units}$$

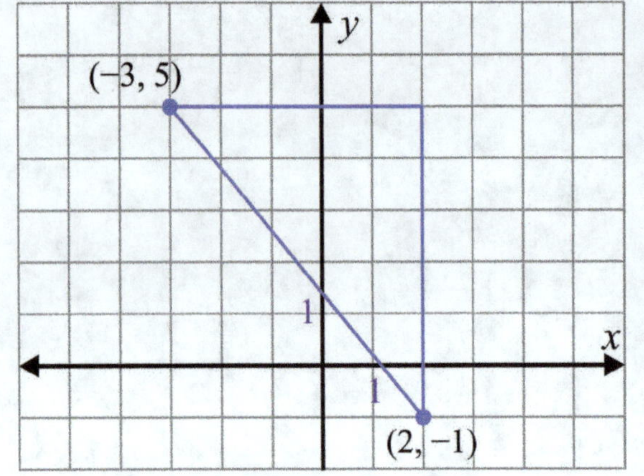

1. Find the distance between the points. Give your answer both in an exact form, and rounded to one decimal digit.

   a. (4, 7) and (−5, 2)

   b. (4, −3) and (−7, 0)

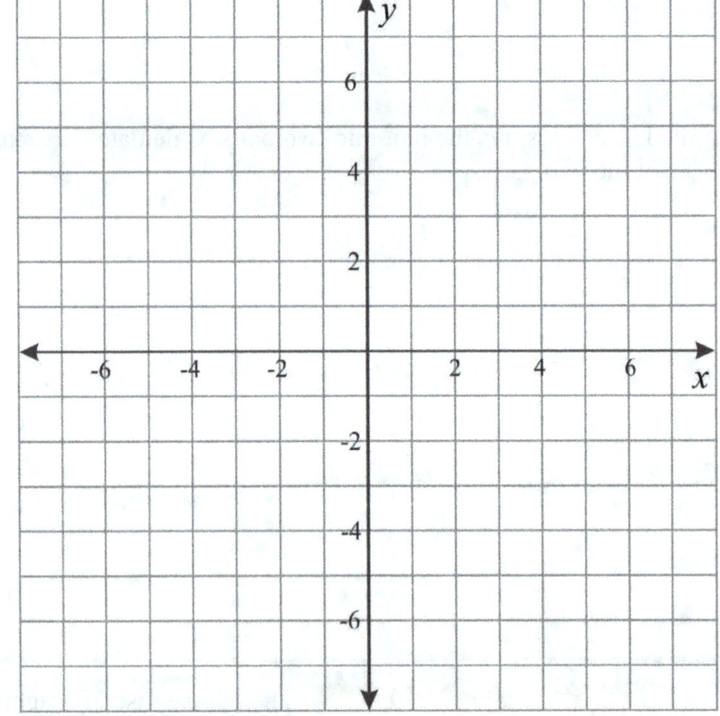

2. Find the distance between (−2, 0) and (8, 11) without drawing the points on the grid. Give your answer both in an exact form, and rounded to one decimal digit.

3. Explain how to find the distance between (5, 2) and (30, 45), and also find the distance.
Hint: consider the horizontal distance and the vertical distance between the points.
Give your answer in an exact form.

4. Find the distances between the points using the Pythagorean Theorem. Give your answer in an exact form.

a. (−10, 9) and (22, 15)	**b.** (30, −25) and (−7, −32)

5. Find the perimeter of the triangle with vertices at (−4, −4), (−4, 5), and (2, 5), to the nearest tenth of a unit.

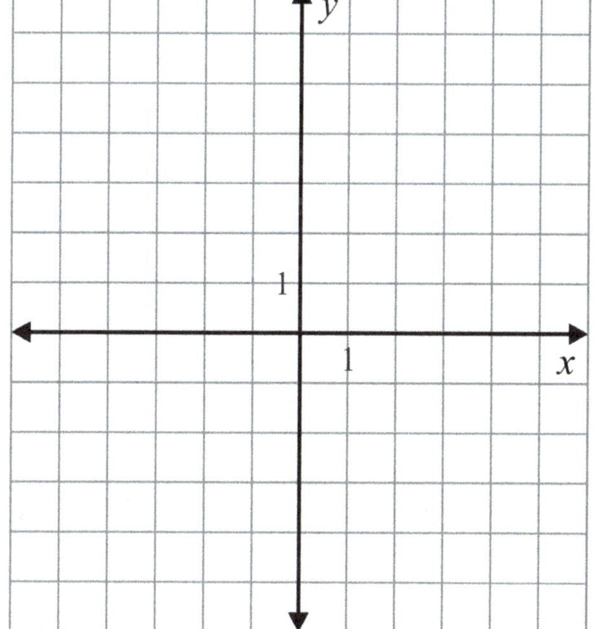

6. Find the area of the square with vertices at (−5, 0), (0, 5), (5, 0), and (0, −5) to the nearest tenth of a square unit.

7. Which shape has a shorter perimeter? How much shorter (to the nearest tenth of a unit)?

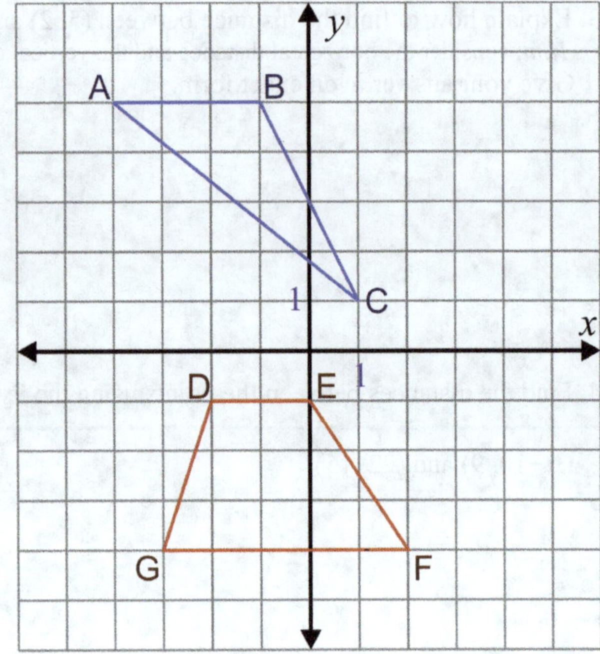

8. How much bigger is the area of trapezoid DEFG than the area of triangle ABC?

9. On a map, Sarah's home is at the origin, and the local swimming pool is at (50, 32). To go there, Sarah has to follow the roads, and travel first to the point (50, 0), then to (50, 32), whereas a crow can fly from Sarah's home directly to the pool. If each unit on the map is 10 meters, how much shorter is the way a crow flies than the way Sarah has to travel?

Puzzle Corner

Find the area of a regular octagon with sides 2 units long.

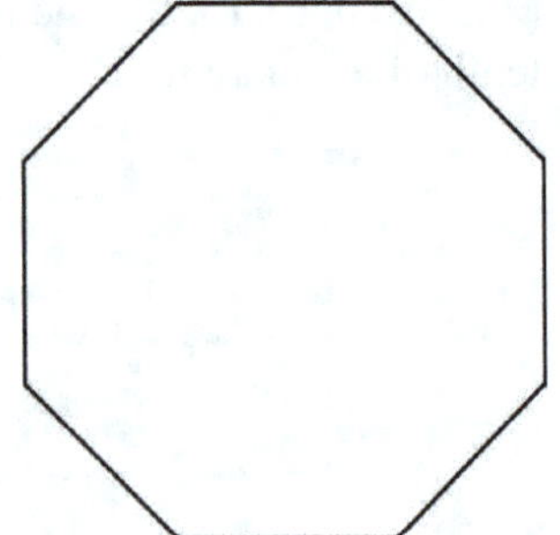

Chapter 6 Review

1. Find the values of these (principal) square roots and cube roots.

a. $\sqrt{64}$	**b.** $\sqrt{169}$	**c.** $\sqrt{2{,}500}$	**d.** $\sqrt{0.81}$
e. $\sqrt{\dfrac{36}{100}}$	**f.** $-\sqrt{49}$	**g.** $\sqrt[3]{125}$	**h.** $\sqrt[3]{27{,}000}$

2. Between which two whole numbers do the following square roots lie? Do not use a calculator.

a. $\sqrt{7}$ **b.** $\sqrt{77}$ **c.** $\sqrt{134}$ **d.** $\sqrt{48}$

3. Find the value of $\sqrt{13}$ to one decimal digit, without a calculator.

4. Plot the following numbers *approximately* on the number line. Do not use a calculator, but think between which two integers the square root lies, and whether it is close to one of those integers, using mental math.

$\sqrt{50}/2$ $\sqrt{17}$ $-\sqrt{8}$ $\sqrt[3]{8}$ $\sqrt{101} - 4$ $-\sqrt[3]{27}$ π $-\sqrt{38}/3$

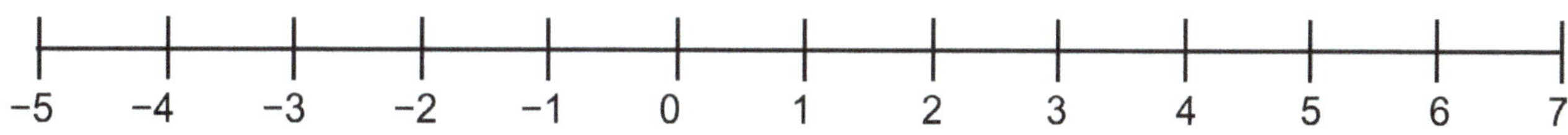

5. Compare, writing >, <, or = between the numbers. Think between which two whole numbers the square root lies, using mental math.

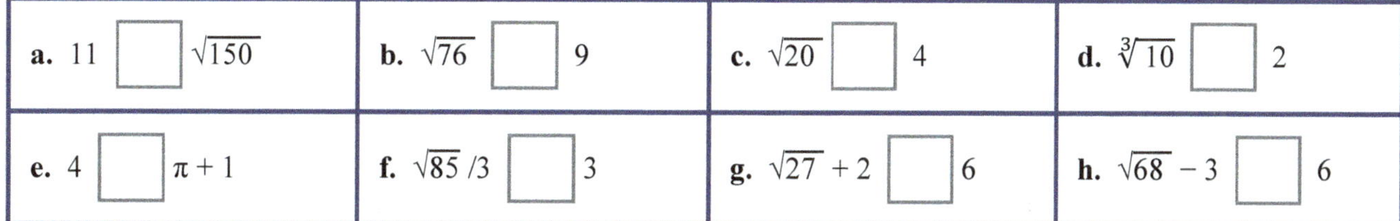

a. 11 ☐ $\sqrt{150}$	**b.** $\sqrt{76}$ ☐ 9	**c.** $\sqrt{20}$ ☐ 4	**d.** $\sqrt[3]{10}$ ☐ 2
e. 4 ☐ $\pi + 1$	**f.** $\sqrt{85}/3$ ☐ 3	**g.** $\sqrt{27} + 2$ ☐ 6	**h.** $\sqrt{68} - 3$ ☐ 6

6. Find the value of the expressions.

a. $\sqrt{144}$	**b.** $-\sqrt{81}$	**c.** $\sqrt{1{,}600}$
d. $\sqrt{10^2 - 6^2}$	**e.** $\sqrt{49 \cdot 49}$	**f.** $\sqrt{5 \cdot (83 - 3)}$

7. **a.** If the side of a square measures $\sqrt{7}$, what is its area?

b. What is the perimeter of a square with an area of 20 square units? Give your answer as an exact value (not rounded).

8. Determine whether the following numbers are rational or irrational, and explain why.

a. 0.8053

b. $\sqrt{2{,}500}$

c. 5π

d. $-\sqrt{56}$

e. $-\frac{2}{7}$

f. $\frac{1}{\sqrt{36}}$

g. $2.1\overline{09}$

h. 0.020202

i. $0.20\overline{8}$

j. $4\sqrt{7}$

9. Write each repeating decimal as a fraction.

a. $0.\overline{61}$	**b.** $4.1\overline{7}$

10. Solve. Give the final answers in exact form.

a. $x^2 = 147$	 **b.** $a^2 = 169$	**c.** $w^3 = 0.36$
d. $3x^3 = 21$	**e.** $5b^3 = 625$	**f.** $2a^3 = -16$

11. Solve. Give your answer to the nearest thousandth. You may use a calculator.

 **a.** $y^2 + 18 = 35$	**b.** $0.6h^2 = 4$

12. For each set of lengths, determine whether they form a right triangle.

a. 20, 24, 30

b. 2.6, 1.0, 2.4

13. Solve for the unknown side. Leave your answer in root form if the radicand is not a perfect square.

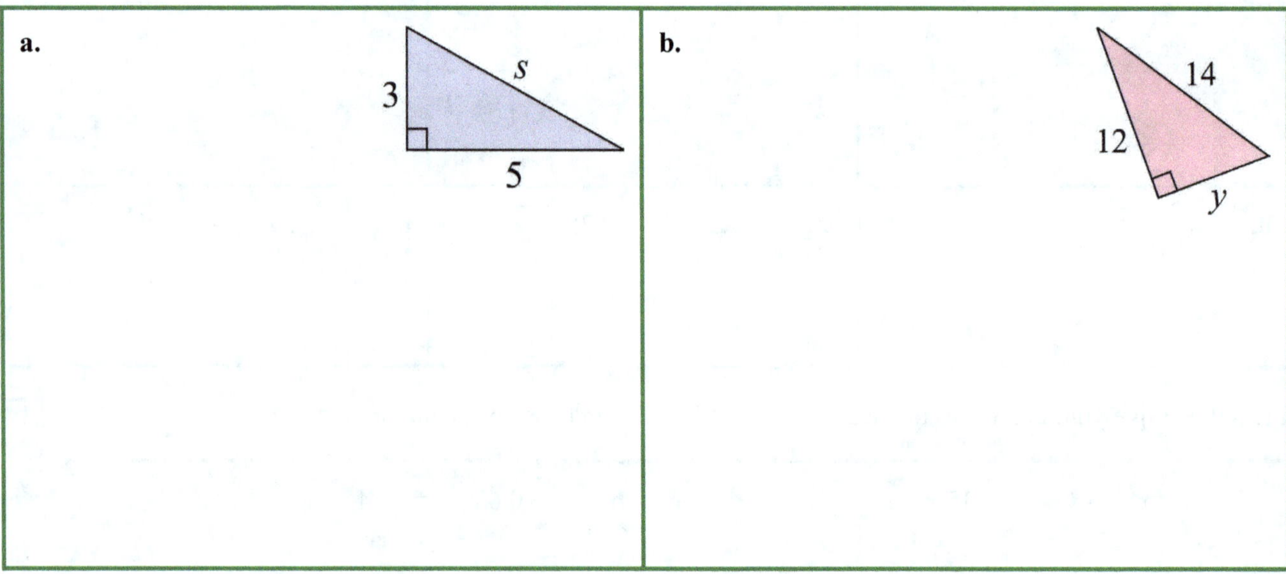

14. Solve for the unknown side. Round your answer to the same accuracy as the given numbers.

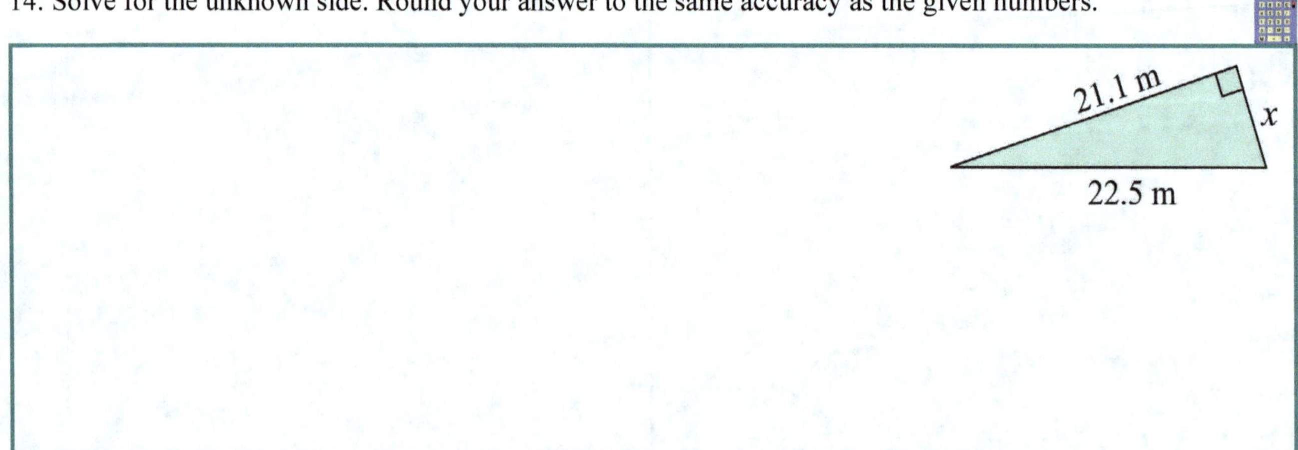

15. The two legs of a right triangle are $\sqrt{7}$ and $\sqrt{8}$.
How long is the hypotenuse?

16. Lauren and Anna want to make this pennant for their jogging club. Calculate its area.

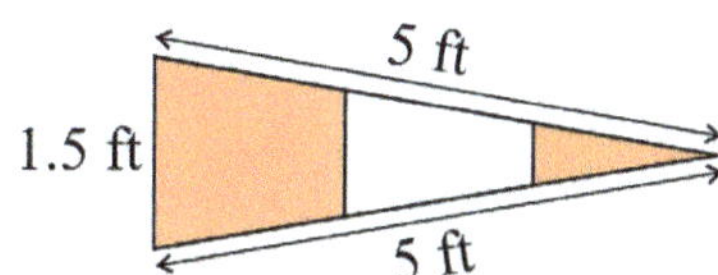

17. Arrange the pieces in the empty square in a manner that will prove the Pythagorean Theorem, and explain how your arrangement does so.

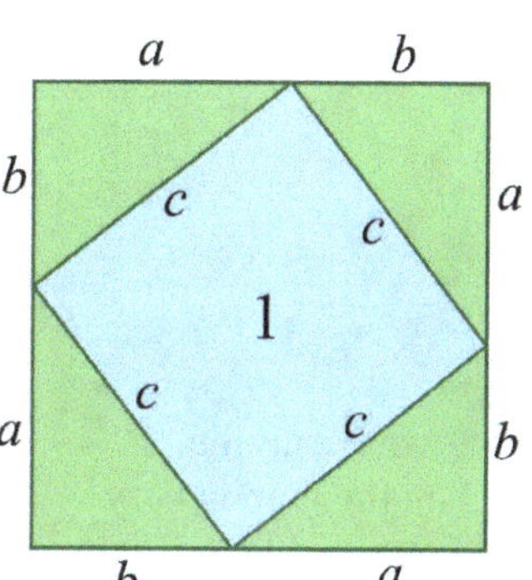

18. Find the area and the perimeter of the garden, if one unit in the grid is 2.0 feet.

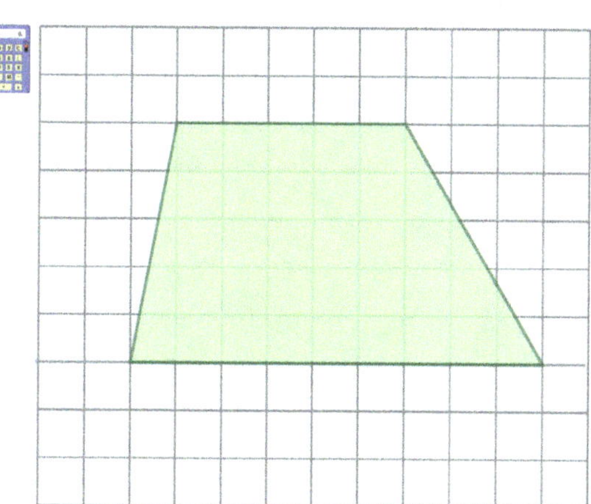

19. Find the distances between the points, to the nearest tenth of a unit.

a. (−3, 4) and (11, −9)	**b.** (42, −15) and (70, 100)

20. Find the diagonal of the cube with edges 2.5 m long.

21. The map shows part of downtown Nashville, Tennessee. The triangle ABC on the map is very close to a right triangle. The distance AB is 370 m and the distance AC is 620 m. However, these distances are approximate, so your calculations will also be only approximate.

About how much shorter is it to travel from point A to point C along Lafayette Street than to travel first along Korean Veterans Boulevard and then along 5th Avenue South?

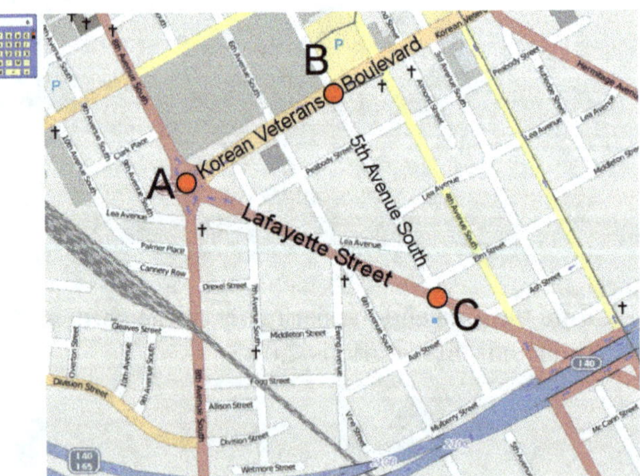

Answers

Square Roots, pp. 7-10

Page 7

1. a. 10 b. 8 c. 2 d. 0
e. 9 f. 12 g. 1 h. 100

2. See the table →

x	x^2
1	1
2	4
3	9
4	16
5	25
6	36
7	49
8	64
9	81
10	100

x	x^2
11	121
12	144
13	169
14	196
15	225
16	256
17	289
18	324
19	361
20	400

3. a. 13 b. 30
c. 15 d. 11
e. 21 f. 90
g. 18 h. 20
i. 80 j. 160
k. 130 l. 1000

Page 8

4. a. Between 2 and 3 b. Between 4 and 5
c. Between 6 and 7 d. Between 9 and 10

5. a. 5 u b. 40 u c. $\sqrt{5}$ u d. $\sqrt{11}$ u

6. a. 8 square units
b. 7
c. $\sqrt{130}$ meters

7. a. 0.4 b. 0.1 c. 1.1
d. 4/5 e. 10/3 or 3 1/3 f. 7/6 or 1 1/6

Page 9

8. a. 8.367 b. 1.732 c. 38.079
d. 0.671 e. 0.913 f. 2.104

9. a. 5 b. 11 c. 8
d. 12 e. 6 f. 5

10. a. 5.191. If your calculator doesn't automatically perform the operations in order, you may need to write down the intermediate results (or enter them into the calculator's memory). If you write them down, keep at least 5 decimal digits. In other words, don't round the intermediate results to 3 decimal digits or your final answer may be off.

b. 59.512

Page 10

11. a. 1,600 cm^2
b. 37 sq. in.

12. a. Check the student's square. The side of the square is about $\sqrt{18}$ cm ≈ 4.2 cm.

b. $4 \cdot \sqrt{18}$ cm ≈ 16.97 cm

13. a. Check the student's square. The side of the square is 4.5 cm.
b. $A = (4.5\text{ cm})^2 = 20.25\text{ cm}^2$

14. GIVES THEM SQUARE ROOTS

Puzzle corner:
$19 = \sqrt{(5+5) \times (6+6+6+6+6+5)+6+5}$

How you enter this into a calculator depends. In some calculators, you would first calculate the radicand (what is under the root), and then press the square root button.

In others, you first press the square root button, then enter the rest.

Irrational Numbers, pp. 11-14

Page 11

1. The intermediate guesses by the student(s) will vary. For example:

Low Guess	$(LG)^2$	$(HG)^2$	High Guess
4.35	18.9225	19.0096	4.36
4.357	18.983449	18.992164	4.358
4.358	18.992164	19.000881	4.359
4.3587	18.99826569	18.99913744	4.3588

From this we can see that to three decimal digits, $\sqrt{19}$ is 4.359.

Page 12

2. The intermediate guesses by the student(s) will vary. The tables are just showing one possibility.

a. $\sqrt{7} \approx 2.65$

Low Guess	$(LG)^2$	$(HG)^2$	High Guess
2.6	6.76	7.29	2.7
2.64	6.9696	7.0225	2.65
2.645	6.996025	7.0225	2.65

Here is also an explanation in words of how this process could possibly go.

First we find two consecutive perfect squares so that 7 is between them: $4 < 7 < 9$. From that fact we know that $2 < \sqrt{7} < 3$. Since 7 is closer to 9 than to 4, let's guess that $\sqrt{7} = 2.6$ and check:

$2.6^2 = 6.76$ Too small. Let's guess bigger:

$2.7^2 = 7.29$

The above guesses show us that $\sqrt{7}$ is between 2.6 and 2.7. Now let's guess what the second decimal digit might be:

$2.65^2 = 7.0225$ Too big. Let's guess smaller:

$2.64^2 = 6.9696$

So $\sqrt{7}$ is between 2.64 and 2.65. Now we just need to know whether it would be rounded to 2.64 or 2.65.

$2.645^2 = 6.996025$

This shows us that $\sqrt{7} > 2.645$, so when rounding to two decimal digits, $\sqrt{7} \approx 2.65$.

b. $\sqrt{51} \approx 7.14$

Low Guess	$(LG)^2$	$(HG)^2$	High Guess
7.1	50.41	51.84	7.2
7.14	50.9796	51.1225	7.15
7.141	50.993881	51.051025	7.145

Or, in words:

First we find two consecutive perfect squares so that 51 is between them: $49 < 51 < 64$. From that fact we know that $7 < \sqrt{51} < 8$. Also, since 51 is much closer to 49 than to 64, $\sqrt{51}$ is much closer to 7 than to 8. Let's first guess that $\sqrt{51} = 7.1$ and go on from there:

$7.1^2 = 50.41$ Too small. Let's guess bigger.

$7.2^2 = 51.84$

So $\sqrt{51}$ is between 7.1 and 7.2. Also, it is closer to 7.1 than to 7.2, because 50.41 is closer to 51 than 51.84 is.

$7.13^2 = 50.8369$ Too small. Let's guess bigger.

$7.14^2 = 50.9796$ Still too small. Let's guess bigger.

$7.15^2 = 51.1225$

So $\sqrt{51}$ is between 7.14 and 7.15. Now we just need to know whether it should be rounded to 7.14 or 7.15.

$7.145^2 = 51.051025$

This shows us that $\sqrt{51} < 7.145$, so when rounding to two decimal digits, $\sqrt{51} \approx 7.14$.

c. $\sqrt{99} \approx 9.95$

Low Guess	$(LG)^2$	$(HG)^2$	High Guess
9.9	98.01	99.0025	9.95
9.94	98.8036	99.0025	9.95
9.945	98.903025	99.0025	9.95

Irrational Numbers, cont.

Page 12

3. No, she is not correct. If we square 3.317, we get $3.317^2 = 11.002489$. This is not exactly 11.

4. If we square 71/50, we will get the fraction $(71^2/50^2) = (71^2/2{,}500)$. Since 71^2 does not equal 5,000, the simplified form of this fraction does not equal 2. Thus, the square root of 2 cannot equal 71/50.

Page 13

5. a. Rational; it can be written as the fraction 928/1000.
   b. Irrational, since 128 is not a perfect square.
   c. Rational, since it is a repeating decimal.
   d. Rational, since it is a terminating decimal.
   e. Irrational. Pi is irrational, and an irrational number divided by a rational number is irrational.
   f. Rational; this equals 10.
   g. Rational; it is a repeating decimal.
   h. Irrational. $\sqrt{3}$ is irrational, and an irrational number multiplied by a rational number is irrational.
   i. Irrational. $\sqrt{15}$ is irrational, and an irrational number divided by a rational number is irrational.
   j. Rational, as it equals 5/2.

Page 14

6. a. Incorrect. 1.272727 is indeed rational, but the reason it is rational is because the decimal expansion terminates. As a fraction, this is 1,272,727/1,000,000.
   b. Incorrect. $\sqrt{49} = 7$ so the entire expression equals 21 and is rational.
   c. Incorrect. $\pi/3$ is irrational, because an irrational number (π) divided by a rational number (3) is irrational.
   d. Correct.

7.

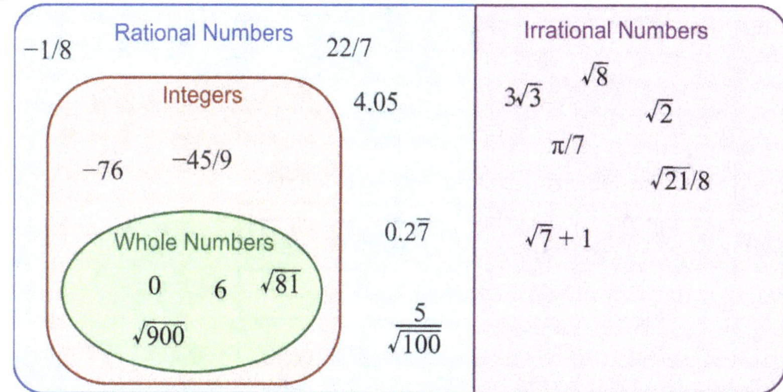

Puzzle corner. Proof. Let x be an irrational and r a rational number. We need to prove that x/r is irrational. Suppose the contrary, that $x/r = s$ is a rational number. Then, $x = rs$, where both r and s are rational, thus are fractions. When two fractions are multiplied, the result is a fraction, or a rational number. So, x would be rational. This is a contradiction. Thus, the original statement is true: x/r is irrational.

Cube Roots and Approximations of Irrational Numbers, pp. 15-18

Page 15

1. a. 3   b. 5   c. 4   d. 10
   e. 1   f. 6   g. 30   h. −2
   i. −1   j. −5   k. 0   l. −20

2. a. 6 cm
   b. 4
   c. 125,000 cm^3
   d. Its edge is 9 in. The surface area is $6 \cdot (9 \text{ in})^2 = 486 \text{ in}^2$.

3. a. 0.2   b. 0.5   c. −0.3
   d. 2/5   e. 4/3   f. −1/2

Cube Roots and Approximations of Irrational Numbers, pp. 15-18

Page 16

4.

a. $5 < \sqrt{31} < 6$	b. $8 < \sqrt{65} < 9$	c. $9 < \sqrt{87} < 10$
d. $-3 < -\sqrt{5} < -2$	e. $-7 < -\sqrt{44} < -6$	f. $-8 < -\sqrt{50} < -7$
g. $1 < \sqrt[3]{7} < 2$	h. $3 < \sqrt[3]{37} < 4$	i. $4 < \sqrt[3]{101} < 5$

5.

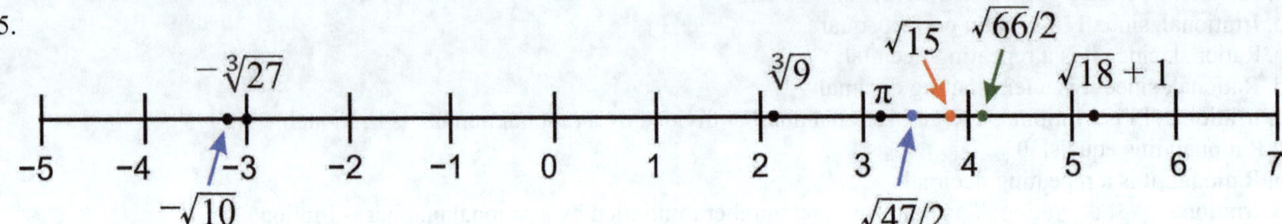

6.

a. $5 < \sqrt{27}$	b. $\sqrt{48} < 7$	c. $\sqrt{18} > 4$	d. $\sqrt[3]{9} > 2$
e. $2 < \sqrt{2} + 1$	f. $\sqrt{32} + 1 > 6$	g. $\sqrt{43} + 5 > 10$	h. $\sqrt{88} - 3 < 7$

7. a. $\sqrt{30}$ lies between 5 and 6, and $\sqrt{60}$ between 7 and 8.

b. Therefore, $2\sqrt{30}$ is between 10 and 12, whereas $\sqrt{2 \cdot 30} = \sqrt{60}$ is between 7 and 8. So, $2\sqrt{30}$ is not equal to $\sqrt{2 \cdot 30}$.

Page 17

8. No, it is not. $\sqrt{50}$ is a little over 7, so $\dfrac{\sqrt{50}}{2}$ is a little over 3.5. On the other hand, $\sqrt{\dfrac{50}{2}} = \sqrt{25} = 5$. So, they are not equal.

9. a. 7.1 b. 9.9 c. 0.8 d. −2.6

10. a. The intermediate guesses by the student(s) will vary. The table is just showing one example.

$\sqrt{11} \approx 3.3$

Low Guess	$(LG)^2$	$(HG)^2$	High Guess
3.2	10.24	11.56	3.4
3.3	10.89	11.2225	3.35
3.31	10.9561	11.0224	3.32

10. b. $\sqrt{11} - \sqrt{2} \approx 3.3 - 1.4 = \underline{1.9}$. $3\sqrt{11} \approx \underline{9.9}$

11. She would take numbers between 6.7 and 6.8 and square those. A good starting point is the midpoint, 6.75. Since $6.75^2 = 45.5625$, a good point for the next guess is something quite a bit less than 6.75, such as 6.72. Let's continue:

$\sqrt{45} \approx 6.71$

Low Guess	$(LG)^2$	$(HG)^2$	High Guess
6.72	45.1584	45.5625	6.75
6.705	44.957025	45.0241	6.71

Page 18

12. Below each irrational number is an approximation for it.

$\sqrt[3]{1}$	<	$\sqrt{5} - 1$	<	$\sqrt[3]{8}$	<	$\sqrt{19}/2$	<	$\sqrt{9}$	<	$\sqrt{13}$	<	$\sqrt[3]{100}$	<	$\sqrt{22} + 1$	<	2π
1		between 1 and 2		2		between 2 and 3		3		between 3 and 4		between 4 and 5		between 5 and 6		6.28

Cube Roots and Approximations of Irrational Numbers, cont.

Page 18

13.

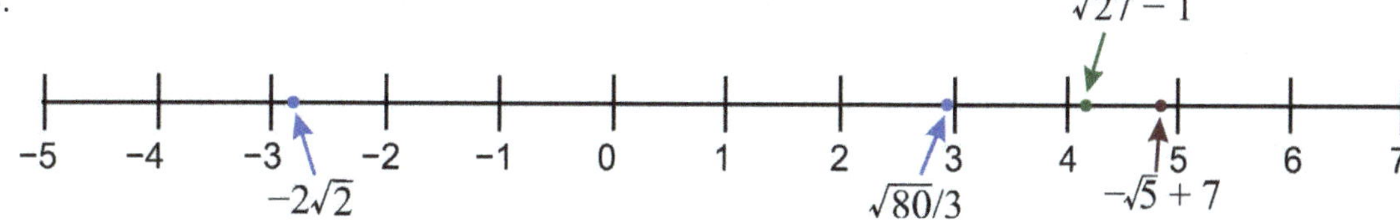

14.

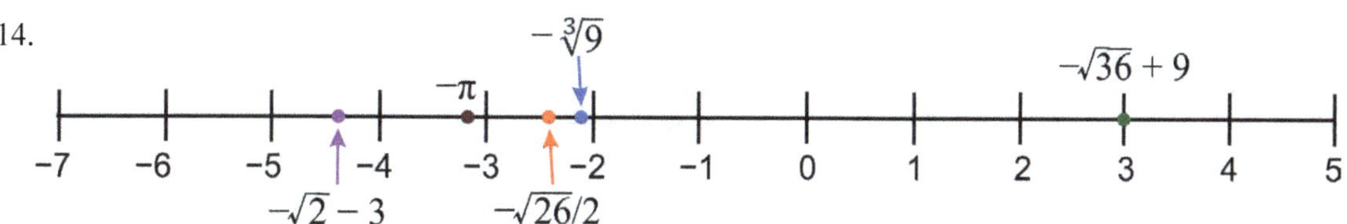

Fractions to Decimals, pp. 19-20

Page 20

1. a. 0.02 b. 0.0278 c. 0.055073
   d. 4.508 e. 5.633 f. 4.0309

2. a. 0.4 b. 0.96 c. 0.27
   d. 1.75 e. 1.32 f. 0.056

3. a. $0.\overline{5}$ b. $18.8\overline{148}$ c. 0.37705

Decimals to Fractions, pp. 21-23

Page 22

1. a. 9382/100 = 4691/50 b. 333/1000 c. 205,056/100,000 = 6408/3125
   d. 61,098/1000 = 30,549/500 e. 45/10,000,000 = 9/2,000,000 f. 4,932,048/1,000,000 = 308,253/62,500

2. No. Their difference is $0.000\overline{2}$.

3.

a. $10x = 4.444...$ $-\ x = 0.444...$ $9x = 4$ $x = \underline{4/9}$	b. $10x = 2.11111...$ $-\ x = 0.21111...$ $9x = 1.9$ $x = 1.9/9 = \underline{19/90}$
c. $1000x = 954.954954...$ $-\ x = 0.954954...$ $999x = 954$ $x = 954/999 = \underline{106/111}$	d. $100x = 253.2323232...$ $-\ x = 2.5323232...$ $99x = 250.7$ $x = 250.7/99 = \underline{2507/990}$

4.

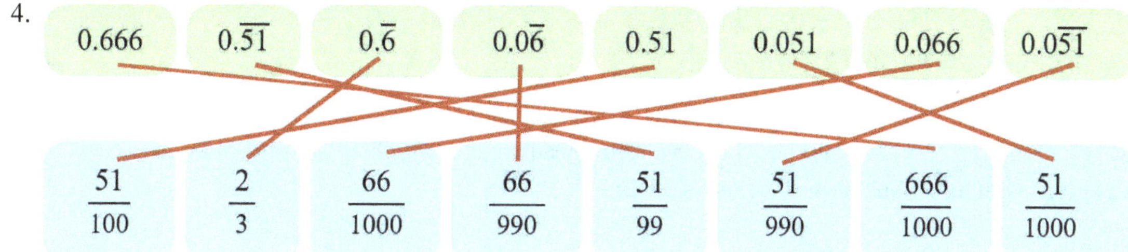

Decimals to Fractions, cont.

Page 23

5. Both are correct. Erica's method is better since it provides a smaller numerator and denominator. Eric's answer is correct but would need simplified.

6. No. It is a fraction so it is rational. The decimal expansion of 305/55494 has 1541 repeating digits, starting at the hundredths place, and the calculator is not able to show that many.

7.

a. $1000x = 256.256256...$ $-\ x = 0.256256...$ $999x = 256$ $x = 256/999$	b. $100x = 305.94949494...$ $-\ x = 3.05949494...$ $99x = 302.89$ $x = 302.89/99 = \underline{30,289/9900}$
c. $10,000x = 321.99\overline{2199}$ $-\ x = 0.03\overline{2199}$ $9999x = 321.96$ $x = 321.96/9999$ $= 32,196/999,900$ The student is not required to simplify this fraction, but it does simplify to 2,683/83,325.	d. $100,000x = 136,309.\overline{36309}$ $-\ x = 1.\overline{36309}$ $99,999x = 136,308$ $x = 136,308/99,999$ $= 45,436/33,333$

Square and Cube Roots as Solutions to Equations, pp. 24-26

Page 24

1.

a. $x^2 = 25$ $x = 5$ or $x = -5$	b. $y^2 = 3,600$ $y = 60$ or $y = -60$
c. $x^2 = 500$ $x = \sqrt{500} \approx 22.36$ or $x = -\sqrt{500} \approx -22.36$	d. $z^2 = 11$ $z = \sqrt{11} \approx 3.32$ or $z = -\sqrt{11} \approx -3.32$
e. $w^2 = 287$ $w = \sqrt{287} \approx 16.94$ or $w = -\sqrt{287} \approx -16.94$	f. $q^2 = 1,000,000$ $q = 1,000$ or $q = -1,000$

Page 25

2. a. $x = 4$ b. $n = 6$ c. $z = 30$
d. $x = 1.91$ e. $b = 4.78$ f. $a = 2.62$

3. a. $s = 8.0$ m b. $s = 29.0$ cm c. $s = 1.80$ ft

4. The edge of that cube is about 2.4216 meters (we will keep a few extra decimals in this intermediate result). Its surface area is $6 \cdot (2.4216\text{ m})^2 = 35.18487936\text{ m}^2$. This last result needs rounded to three significant digits, since the volume was given with three. So, the surface are is 35.2 m^2.

Page 26

5.

a.

$$\begin{aligned} a^2 - 8 &= 37 \\ a^2 &= 45 \\ a &= \sqrt{45} \\ \text{or } a &= -\sqrt{45} \end{aligned}$$

Check:

$$\begin{aligned} (\sqrt{45})^2 - 8 &\stackrel{?}{=} 37 \\ 45 - 8 &\stackrel{?}{=} 37 \\ 37 &= 37 \ \checkmark \end{aligned}$$

b.

$$\begin{aligned} y^2 + 100 &= 1{,}000 \\ y^2 &= 900 \\ y &= 30 \\ \text{or } y &= -30 \end{aligned}$$

Check:

$$\begin{aligned} 30^2 + 100 &\stackrel{?}{=} 1{,}000 \\ 900 + 100 &\stackrel{?}{=} 1{,}000 \\ 1{,}000 &= 1{,}000 \ \checkmark \end{aligned}$$

c.

$$\begin{aligned} b^2 + 1.5 &= 6.4 \\ b^2 &= 4.9 \\ b &= \sqrt{4.9} \\ \text{or } a &= -\sqrt{4.9} \end{aligned}$$

Check:

$$\begin{aligned} (\sqrt{4.9})^2 + 1.5 &\stackrel{?}{=} 6.4 \\ 4.9 + 1.5 &= 6.4 \ \checkmark \end{aligned}$$

d.

$$\begin{aligned} x^2 - 26 &= 709 \\ x^2 &= 735 \\ x &= \sqrt{735} \\ \text{or } y &= -\sqrt{735} \end{aligned}$$

Check:

$$\begin{aligned} (\sqrt{735})^2 - 26 &\stackrel{?}{=} 709 \\ 735 - 26 &= 709 \ \checkmark \end{aligned}$$

6.

a.

$$\begin{aligned} x^3 - 5 &= 59 \\ x^3 &= 64 \\ x &= 4 \end{aligned}$$

Check:

$$\begin{aligned} 4^3 - 5 &\stackrel{?}{=} 59 \\ 64 - 5 &= 59 \ \checkmark \end{aligned}$$

b.

$$\begin{aligned} x^3 + 78 &= 437 \\ x^3 &= 359 \\ x &= \sqrt[3]{359} \end{aligned}$$

Check:

$$\begin{aligned} (\sqrt[3]{359})^3 + 78 &\stackrel{?}{=} 437 \\ 359 + 78 &= 437 \ \checkmark \end{aligned}$$

More Equations that Involve Roots, pp. 27-29

Page 27

1.

a.	b.
$5x^2 = 125$ $x^2 = 25$ $x = 5$ or $x = -5$ Check: $5 \cdot 5^2 \stackrel{?}{=} 125$ $5 \cdot 25 \stackrel{?}{=} 125$ $125 = 125$ ✓	$8.2b^2 = 319$ $b^2 = 319/8.2$ $b = \sqrt{319/8.2} \approx 6.237$ or $b = -\sqrt{319/8.2} \approx -6.237$ Check: $8.2 \cdot 6.237^2 \stackrel{?}{=} 319$ $8.2 \cdot 38.900169 \stackrel{?}{=} 319$ $318.9813858 \approx 319$ ✓
c.	d.
$a^2 + 4.5 = 10.7$ $a^2 = 6.2$ $a = \sqrt{6.2} \approx 2.490$ or $a = -\sqrt{6.2} \approx -2.490$ Check: $(\sqrt{6.2})^2 + 4.5 \stackrel{?}{=} 10.7$ $6.2 + 4.5 \stackrel{?}{=} 10.7$ $10.7 = 10.7$ ✓	$12b^2 = 36{,}000$ $b^2 = 3{,}000$ $b = \sqrt{3{,}000} \approx 54.772$ or $b = -\sqrt{3{,}000} \approx -54.772$ Check: $12 \cdot (\sqrt{3{,}000})^2 \stackrel{?}{=} 36{,}000$ $12 \cdot 3{,}000 \stackrel{?}{=} 36{,}000$ $36{,}000 \approx 36{,}000$ ✓

Page 28

2.

a.	b.
$a^2 + 3^2 = 7^2$ $a^2 + 9 = 49$ $a^2 = 40$ $a = \sqrt{40}$ or $a = -\sqrt{40}$ Check: $(\sqrt{40})^2 + 3^2 \stackrel{?}{=} 7^2$ $40 + 9 \stackrel{?}{=} 49$ $49 = 49$ ✓	$43^2 + x^2 = 51^2$ $x^2 = 51^2 - 43^2$ $x^2 = 752$ $x = \sqrt{752}$ or $x = -\sqrt{752}$ Check: $43^2 + (\sqrt{752})^2 \stackrel{?}{=} 51^2$ $1{,}849 + 752 \stackrel{?}{=} 2{,}601$ $2{,}601 = 2{,}601$ ✓

More Equations that Involve Roots, cont.

Page 28

3.

a.	b.
$s^2 = 2.1^2 + 5.4^2$ $s^2 = 33.57$ $s = \sqrt{33.57} \approx 5.79$ or $s \approx -5.79$ Check: $5.79^2 \stackrel{?}{=} 2.1^2 + 5.4^2$ $33.5241 \approx 33.57$ ✓	$21^2 - w^2 = 15^2$ $-w^2 = 15^2 - 21^2$ $w^2 = 216$ $w = \sqrt{216} \approx 14.70$ or $w = -14.70$ Check: $21^2 - 14.70^2 \stackrel{?}{=} 15^2$ $224.91 \approx 225$ ✓
c.	d.
$121^2 - x^2 = 56$ $-x^2 = 56 - 14{,}641$ $x^2 = 14{,}585$ $x = \sqrt{14{,}585} \approx 120.77$ or $x = -120.77$ Check: $121^2 - 120.77^2 \stackrel{?}{=} 56$ $55.6071 \approx 56$ ✓	$a^2 - 4.5^2 = 5.78$ $a^2 = 5.78 + 20.25$ $a^2 = 26.03$ $a = \sqrt{26.03} \approx 5.10$ or $a = -5.10$ Check: $5.10^2 - 4.5^2 \stackrel{?}{=} 5.78$ $5.76 \approx 5.78$ ✓

Page 29

4.

a.	b.	c.
$56 + s^3 = 542$ $s^3 = 486$ $s = \sqrt[3]{486} \approx 7.862$	$5x^3 = 180$ $x^3 = 36$ $x = \sqrt[3]{36} \approx 3.302$	$254 - z^3 = 46$ $-z^3 = -208$ $z = \sqrt[3]{208} \approx 5.925$

5.

a.	b.	c.
$x^3 = -27$ $x = -3$	$w^3 = -343$ $w = -7$	$4t^3 = -4$ $t^3 = -1$ $t = -1$
d.	e.	f.
$5x^3 + 3 = -27$ $5x^3 = -30$ $x^3 = -6$ $x = -\sqrt[3]{6}$	$-10r^3 = 10{,}000$ $r^3 = -1{,}000$ $r = -10$	$54 - x^3 = -1$ $-x^3 = -55$ $x = \sqrt[3]{55}$

More Equations that Involve Roots, cont.

Page 29

6.

a.	b.	c.
$45 - x^2 = 20$ $-x^2 = -25$ $x^2 = 25$ $x = 5$ or $x = -5$ Check: $45 - 5^2 \stackrel{?}{=} 20$ $45 - 25 \stackrel{?}{=} 20$ $20 = 20$ ✓	$112^2 + s^2 = 18{,}200$ $s^2 = 18{,}200 - 112^2$ $s^2 = 5{,}656$ $s = \sqrt{5{,}656} \approx 75.206$ or $s = -\sqrt{5{,}656} \approx -75.206$ Check: $112^2 + (\sqrt{5{,}656})^2 \stackrel{?}{=} 18{,}200$ $12{,}544 + 5{,}656 \stackrel{?}{=} 18{,}200$ $18{,}200 = 18{,}200$ ✓	$6{,}650 - y^2 = 70^2$ $6{,}650 - 70^2 = y^2$ $1{,}750 = y^2$ $y^2 = 1{,}750$ $y = \sqrt{1{,}750} \approx 41.833$ or $y = -\sqrt{1{,}750} \approx -41.833$ Check: $6{,}650 - (\sqrt{1{,}750})^2 \stackrel{?}{=} 70^2$ $6{,}650 - 1{,}750 \stackrel{?}{=} 4{,}900$ $4{,}900 = 4{,}900$ ✓

Puzzle corner: solve $x^2 - x = 0$. You can use guess and check: Zero fulfills the equation because $0^2 - 0 = 0$. One is also a solution because $1^2 - 1 = 0$.

A way to see the solutions without guessing is to write the equation in the form $x(x - 1) = 0$. The product of x and $x - 1$ can only be zero if either x is zero or $x - 1$ is zero, which means either $x = 0$ or $x = 1$. From this form of the equation we can also see that there are no other solutions.

(We also know that there are no other solutions because of this principle of algebra: an equation where the highest exponent of the variable is n can have at most n solutions within the real numbers. Therefore, our equation, which has 2 as the highest exponent of the variable, can have at most two solutions within the real numbers.)

So the solution is: $x = 0$ or $x = 1$.

The Pythagorean Theorem, pp. 30-34

Page 30

1. $3^2 + 4^2 \stackrel{?}{=} 5^2$

$9 + 16 \stackrel{?}{=} 25$

$25 = 25$

2. Check the student's triangle, so that it has one right angle and the correct measurements. It should have the same shape as this one:

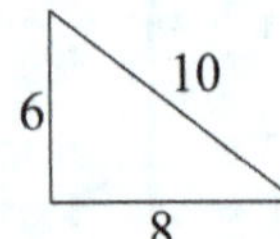

Do the lengths 6, 8, and 10 fulfill the Pythagorean Theorem? Yes. See the calculation below:

$6^2 + 8^2 \stackrel{?}{=} 10^2$

$36 + 64 \stackrel{?}{=} 100$

$100 = 100$

Page 31

3. Since we are solving for a length of side, we ignore the negative roots in these equations.

a. $5^2 + 3^2 = y^2$ $25 + 9 = y^2$ $34 = y^2$ $y = \sqrt{34}$	b. $x^2 + 7^2 = 17^2$ $x^2 + 49 = 289$ $x^2 = 240$ $x = \sqrt{240}$
c. $5^2 + 6^2 = s^2$ $25 + 36 = s^2$ $61 = s^2$ $s = \sqrt{61}$	d. $w^2 + 11^2 = 12^2$ $w^2 + 121 = 144$ $w^2 = 23$ $w = \sqrt{23}$

4. Let d be the length of the diagonal. The diagonal divides the square into two right triangles, each having sides 6, 6, and d. Using the Pythagorean Theorem, we get:

$6^2 + 6^2 = d^2$, from which $d = \sqrt{72} \approx 8.49$. The diagonal measures 8.49 units.

Page 32

5. Since we are solving for a length of side, we ignore the negative roots in these equations.

a. $r^2 + 7^2 = (\sqrt{113})^2$ $r^2 + 49 = 113$ $r^2 = 64$ $r = 8$	b. $x^2 + (\sqrt{52})^2 = 9^2$ $x^2 + 52 = 81$ $x^2 = 29$ $x = \sqrt{29}$
c. $s^2 + (\sqrt{24})^2 = 8^2$ $s^2 + 24 = 64$ $s^2 = 40$ $s = \sqrt{40}$	d. $x^2 + x^2 = (\sqrt{20})^2$ $2x^2 = 20$ $x^2 = 10$ $x = \sqrt{10}$

6. a. Let x be the hypotenuse. We get:

$$(\sqrt{7})^2 + (\sqrt{8})^2 = x^2$$
$$7 + 8 = x^2$$
$$x^2 = 15$$
$$x = \sqrt{15}$$

b. Let s be the unknown leg. We get:

$$s^2 + (\sqrt{41})^2 = (\sqrt{67})^2$$
$$s^2 + 41 = 67$$
$$s^2 = 26$$
$$s = \sqrt{26}$$

Page 33

7. a.

$$w^2 + 37.0^2 = 42.1^2$$
$$w^2 + 1{,}369 = 1{,}772.41$$
$$w^2 = 403.41$$
$$w = \sqrt{403.41} \approx 20.1 \text{ cm}$$

b.

$$t^2 + 1044^2 = 1131^2$$
$$t^2 + 1{,}089{,}936 = 1{,}279{,}161$$
$$t^2 = 189{,}225$$
$$t = \sqrt{189{,}225} = 435 \text{ ft}$$

c.

$$17^2 + 51^2 = x^2$$
$$289 + 2{,}601 = x^2$$
$$x^2 = 2{,}890$$
$$x = \sqrt{2{,}890} \approx 54 \text{ m}$$

Page 33

8. To be able to use the Pythagorean Theorem, we need to convert the lengths of the sides into inches: 12 ft 5 in = 149 in and 7 ft 8 in = 92 in. Let x be the hypotenuse. Then:

$$149^2 + 92^2 = x^2$$
$$22{,}201 + 8{,}464 = x^2$$
$$30{,}665 = x^2$$
$$x = \sqrt{30{,}665} \approx 175.11$$

Since the legs were given to the accuracy of one inch, we will do the same with the hypotenuse. The hypotenuse measures about 175 in or 14 ft 7 in.

Page 34

9. a. Check the student's work. If you are using the digital version, the measurements below will not match your triangle unless the student page was printed at 100% (not "shrink to fit" or similar setting).

The sides measure 67 mm, 63 mm, and between 22 mm and 23 mm. Here, I used 63, 23, and 67:

$$63^2 + 23^2 \stackrel{?}{=} 67^2$$
$$3{,}969 + 529 \stackrel{?}{=} 4{,}489$$
$$4{,}498 \approx 4{,}489$$

While it may seem to you that 4,498 and 4,489 are quite different, they are actually very close to each other. To check how close they are, we must not simply look at their difference of 9, but at the *percentage* difference = (difference/reference). To calculate that, I will use the average value of 4,493.5 as reference.

Percentage difference = (difference/reference) = 9/4,493.5 ≈ 0.0020 = 0.2%. This is an extremely small percentual difference.

10. Let x be the length of the diagonal. Applying the Pythagorean Theorem we get:

$$x^2 = 9.0^2 + 14.4^2$$
$$x^2 = 81 + 207.36$$
$$x^2 = 288.36$$
$$x = \sqrt{288.36} \text{ in} \approx \underline{17.0 \text{ in}}$$

The Pythagorean Theorem, cont.

Page 34

Puzzle corner. The hypotenuse, 108 units, is shorter than one of the legs, 125 units. To fix it, the teacher could switch the two numbers so that the hypotenuse measures 125 units and the leg 108 units.

In that case, we get:

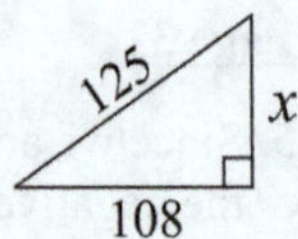

$$\begin{aligned} x^2 + 108^2 &= 125^2 \\ x^2 + 11{,}664 &= 15{,}625 \\ x^2 &= 3{,}961 \\ x &= \sqrt{3{,}961} \text{ units} \approx 62.9 \text{ units} \end{aligned}$$

Applications of The Pythagorean Theorem 1, pp. 35-37

Page 35

1. The side of a square with an area of 100 m^2 is 10 m. The diagonal, d, is given by the Pythagorean Theorem:

$$\begin{aligned} d^2 &= 10^2 + 10^2 \\ d^2 &= 100 + 100 \\ d^2 &= 200 \\ d &= \sqrt{200} \text{ m} \approx \underline{14.1 \text{ m}} \end{aligned}$$

2. The length of the diagonal, d, is given by the Pythagorean Theorem:

$$\begin{aligned} d^2 &= 48^2 + 30^2 \\ d^2 &= 2{,}304 + 900 \\ d^2 &= 3{,}204 \\ d &= \sqrt{3{,}204} \text{ m} \approx 56.6 \text{ m} \end{aligned}$$

The walk around the park is 48 m + 30 m = 78 m. That route is therefore 78 m − 56.6 m = 21.4 m longer.

Page 36

3. We use the right triangle shown in the image.

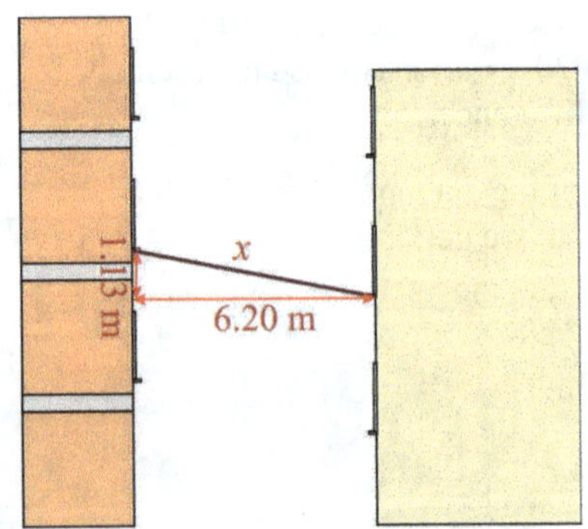

The side 1.13 m comes from subtracting 6.40 m − 5.27 m = 1.13 m. We get:

$$\begin{aligned} x^2 &= 1.13^2 + 6.2^2 \\ x^2 &= 1.2769 + 38.44 \\ x^2 &= 39.7169 \\ x &= \sqrt{39.7169} \text{ m} \approx 6.30 \text{ m} \end{aligned}$$

The clothes line is about 6.30 m long.

4. Let x be the length of AC. Then:

$$\begin{aligned} x^2 &= 6^2 + 3^2 \\ x^2 &= 36 + 9 \\ x^2 &= 45 \\ x &= \sqrt{45} \approx \underline{6.7} \end{aligned}$$

Then, the perimeter is 6.7 + 3 + 6 = 15.7 units.

Page 37

5. To calculate the height, we use the Pythagorean Theorem:

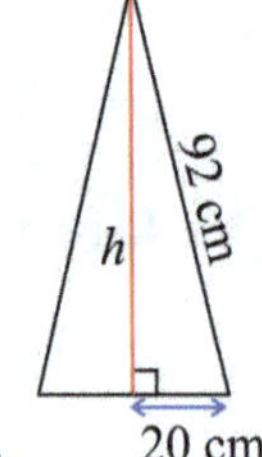

$$\begin{aligned} 20^2 + h^2 &= 92^2 \\ h^2 &= 92^2 - 20^2 \\ h^2 &= 8{,}064 \\ h &= \sqrt{8{,}064} \text{ cm} \approx 89.7998 \text{ cm} \end{aligned}$$

Then, the area is $A = bh/2$ = 89.7998 cm · 40 cm / 2 = 1,795.996 cm^2 ≈ $\underline{1{,}800 \text{ cm}^2}$.

6. First, we calculate the altitude with the Pythagorean Theorem:

24 cm

h

12 cm

$$\begin{aligned} 12^2 + h^2 &= 24^2 \\ h^2 &= 24^2 - 12^2 \\ h^2 &= 432 \\ h &= \sqrt{432} \text{ cm} \approx 20.7846 \text{ cm} \end{aligned}$$

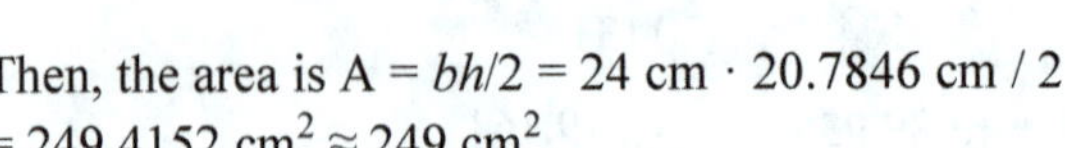

Then, the area is $A = bh/2$ = 24 cm · 20.7846 cm / 2 = 249.4152 cm^2 ≈ $\underline{249 \text{ cm}^2}$.

A Proof of the Pythagorean Theorem and of Its Converse, pp. 38-41

Page 38

1.

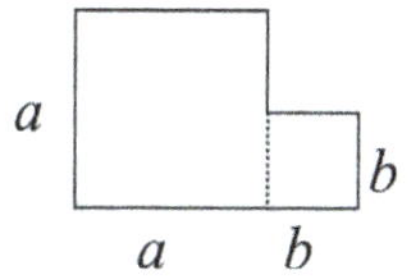

First, we have two squares with areas a^2 and b^2. The total area of the figure is therefore $a^2 + b^2$.

→

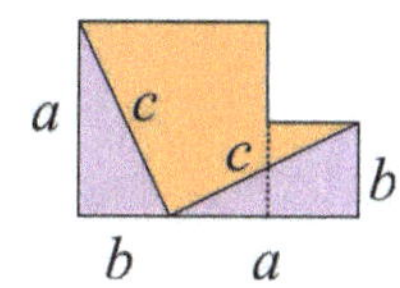

Two lines are drawn so that two right triangles with legs *a* and *b* are formed.

→

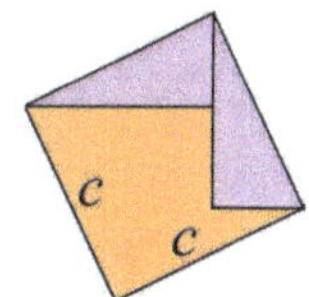

The two right triangles are moved into new positions. Now we have a square with sides *c* units long and an area of c^2.

Since the total area of the figure is preserved through these changes, $a^2 + b^2 = c^2$.

Page 39

3. a. STATEMENT: If there are no clouds on the sky, then it is not raining. <u>true</u> / false

CONVERSE: If it is not raining, then there are no clouds on the sky. true / <u>false</u>

b. STATEMENT: If you eat spoiled food, you will get gastrointestinal symptoms. <u>true</u> / false

CONVERSE: If you get gastrointestinal symptoms, then you eat/ate spoiled food. true / <u>false</u>

c. STATEMENT: If you just had your 7th birthday, then you'll turn 8 in a year. <u>true</u> / false

CONVERSE: If you'll turn 8 in a year, then you just had your 7th birthday. <u>true</u> / false

d. STATEMENT: If $a + b = c$, then $c - b = a$. <u>true</u> / false

CONVERSE: If $c - b = a$, then $a + b = c$. <u>true</u> / false

e. STATEMENT: If you multiply two odd numbers, the product is also odd. <u>true</u> / false

CONVERSE: If the product of two numbers is odd, the numbers that were multiplied are odd. <u>true</u> / false

Page 40

5. We check if the three numbers fulfill the Pythagorean Theorem:

$$40.2^2 + 36.4^2 \stackrel{?}{=} 49.1^2$$
$$1{,}616.04 + 1{,}324.96 \stackrel{?}{=} 2{,}410.81$$
$$2{,}941 \neq 2{,}410.81$$

No, the corner is not a right triangle, as 2,941 is very different from 2,410.81.
(You can check that by calculating the percent relative difference using the average of the two numbers as the reference value. You should get 530.19/2675.905 ≈ 19.81%.)

Page 41

6. They can measure the two diagonals and check that they are equal. If so, the quadrilateral cannot be a (non-rectangular) parallelogram, so it must be a rectangle. (In a parallelogram that is not a rectangle, the diagonals are of different lengths.)

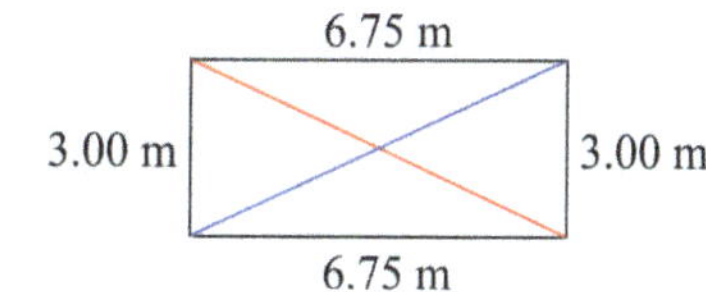

Another possibility is to actually calculate the length of the diagonal, and then measure to check that the measurement agrees with the calculation. The Pythagorean Theorem applied to the triangle gives us:

$$x^2 = 3^2 + 6.75^2$$
$$x^2 = 9 + 45.5625$$
$$x^2 = 54.5625$$
$$x = \sqrt{54.5625}\text{ m} \approx 7.39\text{ m}$$

So if the diagonals measure 7.39 m, the shape is a rectangle.

A Proof of the Pythagorean Theorem and of Its Converse, cont.

Page 41

7.

a. $6^2 + 9^2 \stackrel{?}{=} 13^2$

$36 + 81 \stackrel{?}{=} 169$

$117 < 169$

The triangle is obtuse.

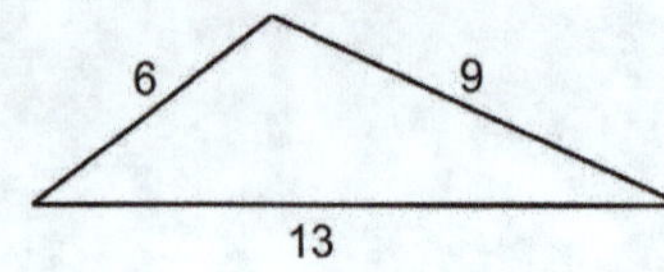

b. $12^2 + 5^2 \stackrel{?}{=} 13^2$

$144 + 25 \stackrel{?}{=} 169$

$169 = 169$

The triangle is right.

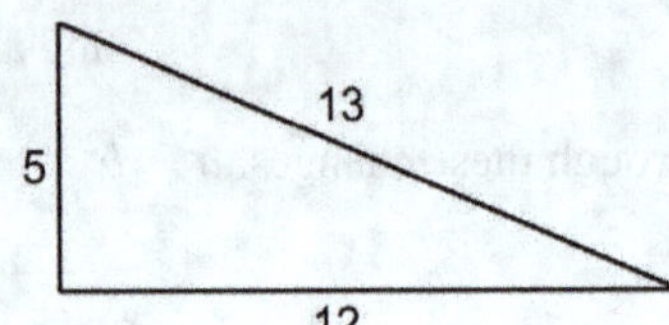

c. $4^2 + 5^2 \stackrel{?}{=} 7^2$

$16 + 25 \stackrel{?}{=} 49$

$41 < 49$

The triangle is obtuse.

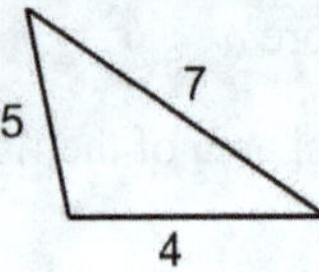

d. $4^2 + 5^2 \stackrel{?}{=} 6^2$

$16 + 25 \stackrel{?}{=} 36$

$41 > 36$

The triangle is acute.

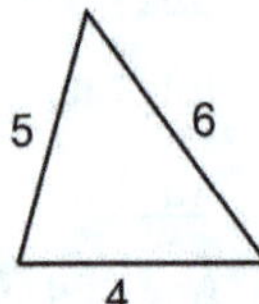

e. $11^2 + 10^2 \stackrel{?}{=} 13^2$

$121 + 100 \stackrel{?}{=} 169$

$221 > 169$

The triangle is acute.

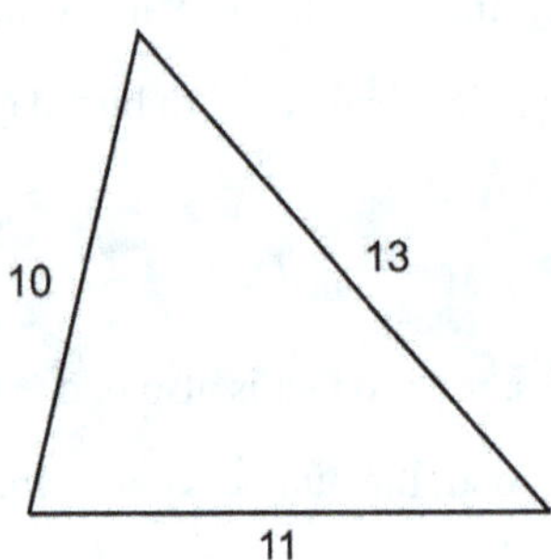

f. $15^2 + 20^2 \stackrel{?}{=} 25^2$

$225 + 400 \stackrel{?}{=} 625$

$625 = 625$

The triangle is right.

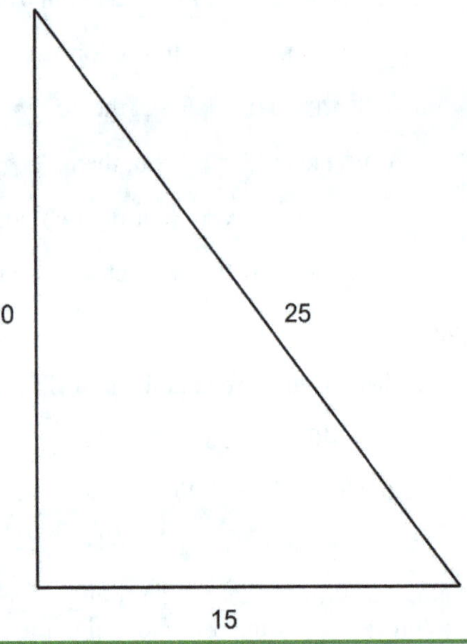

Applications of The Pythagorean Theorem 2, pp. 42-45

Page 42

1. Using the Pythagorean Theorem:

$$(\sqrt{18})^2 + 5^2 = x^2$$
$$18 + 25 = x^2$$
$$43 = x^2$$
$$x = \sqrt{43} \approx 6.56 \text{ units}$$

2. Note the image on the right:

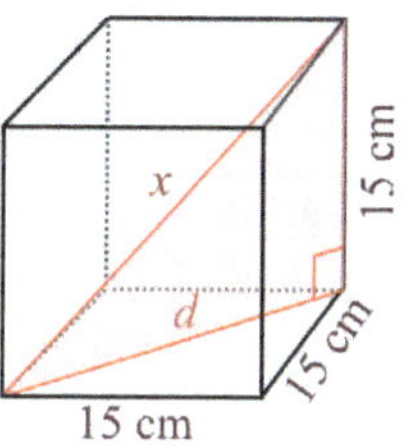

We will first solve for d using the right triangle on the bottom of the cube:

$$15^2 + 15^2 = d^2$$
$$450 = d^2$$
$$d = \sqrt{450}$$

Next, we will use the Pythagorean Theorem again in the pink triangle:

$$(\sqrt{450})^2 + 15^2 = x^2$$
$$450 + 225 = x^2$$
$$x = \sqrt{675}$$
$$x \approx \underline{26 \text{ cm}}$$

Page 43

3. In order to calculate the surface area, we need to find out the height of each of the triangular faces (h in the image).

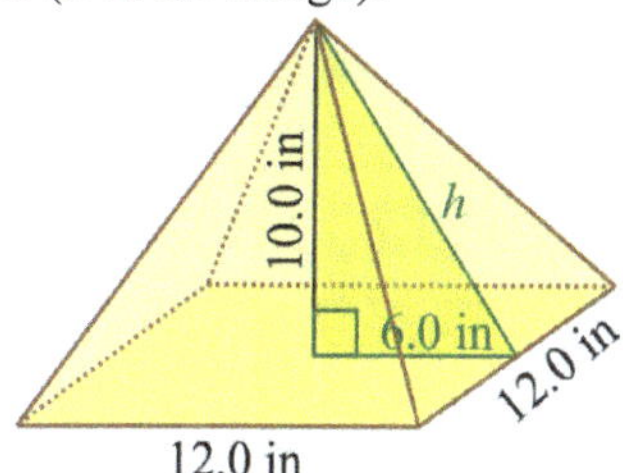

The height of the pyramid (10 in), half of its bottom edge (6 in), and h form a right triangle. Using the Pythagorean Theorem, we get:

$$6^2 + 10^2 = h^2$$
$$36 + 100 = h^2$$
$$h = \sqrt{136}$$

Now, the surface area is the area of the bottom square, plus four times the area of one of the triangular faces:

$A = (12 \text{ in})^2 + 4 \cdot 12 \text{ in} \cdot \sqrt{136} \text{ in} / 2$
$\approx 144 \text{ in}^2 + 279.88569 \text{ in}^2 \approx \underline{424 \text{ in}^2}$.

4. a.

$$rafter^2 = 3^2 + 12^2$$
$$rafter^2 = 9 + 144$$
$$rafter^2 = 153$$
$$rafter = \sqrt{153} \approx 12.37 \text{ ft} \approx \underline{12 \text{ ft } 4 \text{ in}}$$

b. The rise of 5 ft 3 in is 5 1/4 ft = 5.25 ft.

$$rafter^2 = 12^2 + 5.25^2$$
$$rafter^2 = 144 + 27.5625$$
$$rafter^2 = 171.5625$$
$$rafter = \sqrt{171.5625} \text{ ft} \approx 13.10 \text{ ft} \approx \underline{13 \text{ ft } 1 \text{ in}}$$

Alternatively, you could calculate everything in inches (instead of in feet) and lastly convert to feet and inches. The answers will be the same as listed above.

Page 44

5. The roof consists of two identical rectangles. One dimension of each rectangle is given as 5 m. We need to calculate the other using the Pythagorean Theorem in the below triangle:

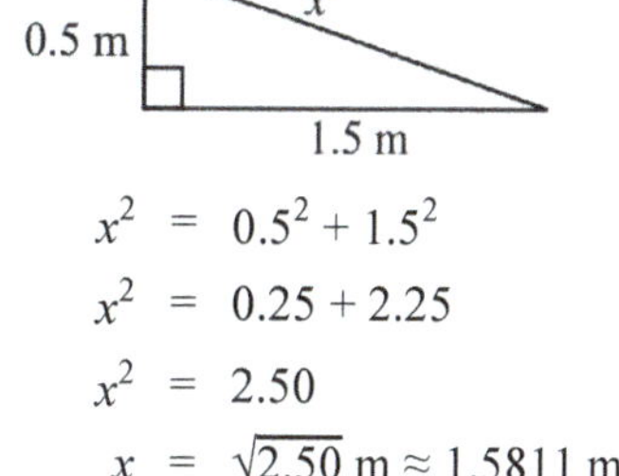

$$x^2 = 0.5^2 + 1.5^2$$
$$x^2 = 0.25 + 2.25$$
$$x^2 = 2.50$$
$$x = \sqrt{2.50} \text{ m} \approx 1.5811 \text{ m}$$

So the area of the roof is $2 \cdot 5 \text{ m} \cdot 1.5811 \text{ m}$
$= 15.811 \text{ m}^2 \approx \underline{15.8 \text{ m}^2}$.

6. The roof consists of four identical isosceles triangles. To calculate the area of those triangles, we need to find the altitude, h, of the triangles. You see one of the triangles below.

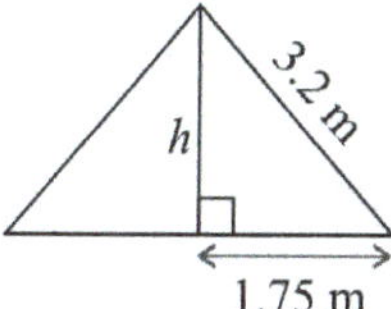

Applying the Pythagorean Theorem, we get:

$$1.75^2 + h^2 = 3.2^2$$
$$h^2 = 3.2^2 - 1.75^2$$
$$h^2 = 7.1775$$
$$h = \sqrt{7.1775} \text{ m} \approx 2.679 \text{ m}$$

The total surface area is then
$4 \cdot 3.5 \text{ m} \cdot 2.679 \text{ m} / 2 \approx \underline{18.8 \text{ m}^2}$.

Applications of The Pythagorean Theorem 2, cont.

Page 45

7. a. We can calculate the length of the creek by applying the Pythagorean Theorem to the right triangle in the image:

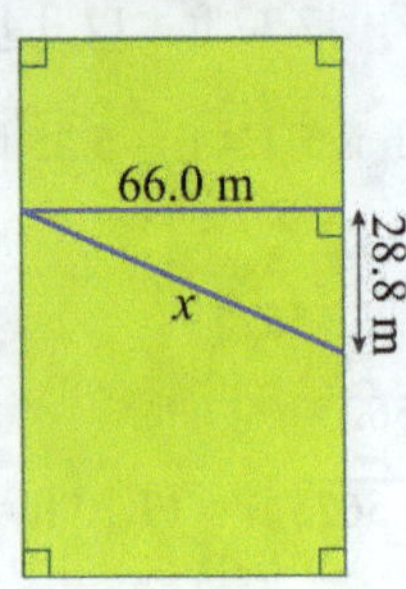

$$x^2 = 66^2 + 28.8^2$$
$$x^2 = 4{,}356 + 829.44$$
$$x^2 = 5{,}185.44$$
$$x = \sqrt{5{,}185.44}\text{ m} \approx 72.0\text{ m}$$

There is also another way to draw a right triangle into the picture, but its dimensions are the same.

7. b. The two areas are trapezoids:

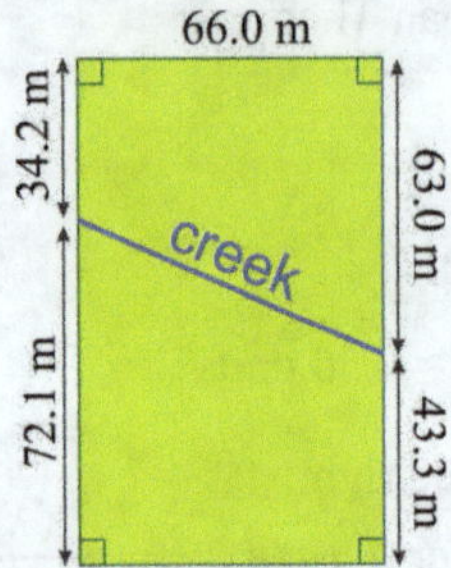

The northern one has a height of 66.0 m, and the two parallel sides measure 34.2 m and 63.0 m. The area is then:

$(34.2\text{ m} + 63.0\text{ m})/2 \cdot 66.0\text{ m} = 3{,}207.6\text{ m}^2 \approx \underline{3{,}210\text{ m}^2}$

Similarly, the area of the southern part is

$(72.1\text{ m} + 43.3\text{ m})/2 \cdot 66.0\text{ m} = 3{,}808.2\text{ m}^2 \approx \underline{3{,}810\text{ m}^2}$

Puzzle corner. Let x be the length of the one leg. Then the other leg is $2x$, and we can write and solve the equation:

$$x^2 + (2x)^2 = (12.0\text{ m})^2$$
$$x^2 + 4x^2 = 144\text{ m}^2$$
$$5x^2 = 144\text{ m}^2$$
$$x^2 = 144/5\text{ m}^2$$
$$x = \sqrt{144/5}\text{ m} \approx 5.37\text{ m}$$

The other leg is $2\sqrt{144/5}\text{ m} \approx 10.73$ m.

The sides of the triangle measure 5.37 m, 10.73 m, and 12.0 m.

Distance Between Points, pp. 46-48

Page 46

1. a. When we connect the points and then draw a right triangle like in Example 1, the sides of the right triangle are <u>9 units</u> and <u>5 units</u> long. Using the Pythagorean Theorem, the desired distance is

$$9^2 + 5^2 = x^2$$
$$81 + 25 = x^2$$
$$106 = x^2$$
$$x = \sqrt{106} \approx 10.3\text{ units}$$

b. The two sides of the right triangle are now 11 units and 3 units, and the equation is:

$$11^2 + 3^2 = x^2$$
$$121 + 9 = x^2$$
$$130 = x^2$$
$$x = \sqrt{130} \approx 11.4\text{ units}$$

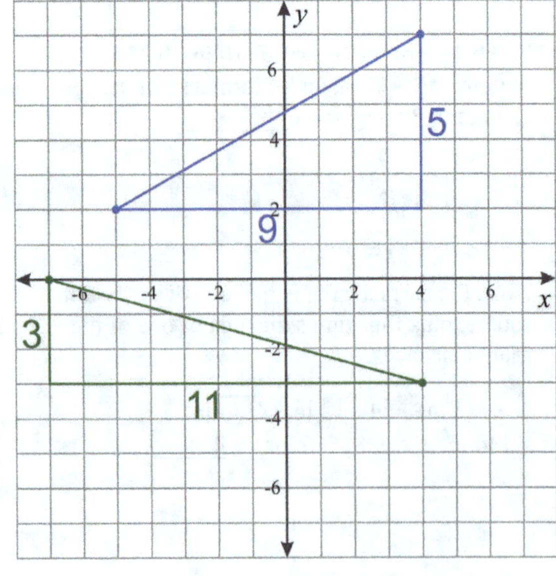

Page 46

2. We need to find the vertical distance and the horizontal distance between the given points. Those will be the two legs of a right triangle that has the line between the two points as the hypotenuse.

   The horizontal distance between the two points is comes from their x-coordinates and is $|-2 - 8|$ = 10 units. The vertical distance comes from the y-coordinates and is $|0 - 11|$ = 11 units long. Using the Pythagorean Theorem, the desired distance between the two points is

$$\begin{aligned} 10^2 + 11^2 &= x^2 \\ 100 + 121 &= x^2 \\ 221 &= x^2 \\ x &= \sqrt{221} \approx 14.9 \text{ units} \end{aligned}$$

Page 47

3. First find the horizontal distance between the two points, which is found by taking the absolute value of the difference of their x-coordinates. Then find the vertical distance between the two points by taking the absolute value of the difference of their y-coordinates. Use those two distances as the legs of a right triangle, and solve for the hypotenuse using the Pythagorean Theorem.

   Here, the horizontal distance is $|30 - 5|$ = 25 units. The vertical distance is $|45 - 2|$ = 43 units. Using the Pythagorean Theorem, we get:

$$\begin{aligned} 25^2 + 43^2 &= x^2 \\ 625 + 1{,}849 &= x^2 \\ 2{,}474 &= x^2 \\ x &= \sqrt{2{,}474} \text{ units} \end{aligned}$$

4.

a. (−10, 9) and (22, 15)

The horizontal distance is $|22 - (-10)|$ = 32 units. The vertical distance is $|15 - 9|$ = 6 units.

$$\begin{aligned} 32^2 + 6^2 &= x^2 \\ 1024 + 36 &= x^2 \\ x &= \sqrt{1{,}060} \text{ units} \end{aligned}$$

b. (30, −25) and (−7, −32)

The horizontal distance is $|-7 - 30|$ = 37 units. The vertical distance is $|-32 - (-25)|$ = 7 units.

$$\begin{aligned} 37^2 + 7^2 &= x^2 \\ 1369 + 49 &= x^2 \\ x &= \sqrt{1{,}418} \text{ units} \end{aligned}$$

5. The distance between (−4, −4) and (−4, 5) is simply 9 units (it's their vertical distance). The distance between (−4, 5) and (2, 5) is simply 6 units (it's their horizontal distance). Let x be the distance between (−4, −4) and (2, 5). Then,

$$\begin{aligned} x^2 &= (-4 - 2)^2 + (-4 - 5)^2 \\ x^2 &= (-6)^2 + (-9)^2 \\ x^2 &= 36 + 81 \\ x &= \sqrt{117} \text{ units} \approx 10.8 \text{ units} \end{aligned}$$

   The perimeter is then 10.8 + 9 + 6 = 25.8 units.

6. First we find the side of the square. Let s be the distance between (5, 0) and (0, 5). Then,

$$\begin{aligned} s^2 &= 5^2 + 5^2 \\ s^2 &= 25 + 25 \\ s &= \sqrt{50} \end{aligned}$$

   The area of the square is then $s^2 = (\sqrt{50})^2$ = 50 square units.

Page 48

7. In triangle ABC, the distance from A to C, or AC, is $\sqrt{4^2 + 5^2}$ units = $\sqrt{41}$ units.
   The distance from B to C, or BC is $\sqrt{2^2 + 4^2}$ units = $\sqrt{20}$ units. And AB is 3 units.

   The perimeter is $\sqrt{41} + \sqrt{20} + 3 \approx 13.875$ units

   In trapezoid DEGF, DE = 2 units, GF = 5 units, EF = $\sqrt{2^2 + 3^2} = \sqrt{13}$ units, and DG = $\sqrt{1^2 + 3^2} = \sqrt{10}$ units. The perimeter is $\sqrt{13} + \sqrt{10} + 2 + 5 \approx 13.768$ units.

   The trapezoid has a shorter perimeter, by 13.875 − 13.768 ≈ 0.1 units.

8. If AB = 3 units is the base of the triangle, then its height is 4 units, and its area is 3 · 4 ÷ 2 = 6 square units.

   The area of the trapezoid is the average of its two parallel sides times its height. The two parallel sides measure 2 and 5 units, so their average is 3.5 units. Its height is 3 units. So, its area is 3 · 3.5 = 10.5 square units.

   The area of the trapezoid is 4.5 square units bigger than the area of the triangle.

Distance Between Points, Cont.

Page 48

9. First, let's calculate the distances using units on the map. Sarah has to travel 50 + 32 = 82 map units. The crow flies a distance of $\sqrt{50^2 + 32^2} = \sqrt{3524} \approx 59.363$ map units. The difference between what Sarah travels and what the crow flies is about 22.6 map units, which is about 226 meters.

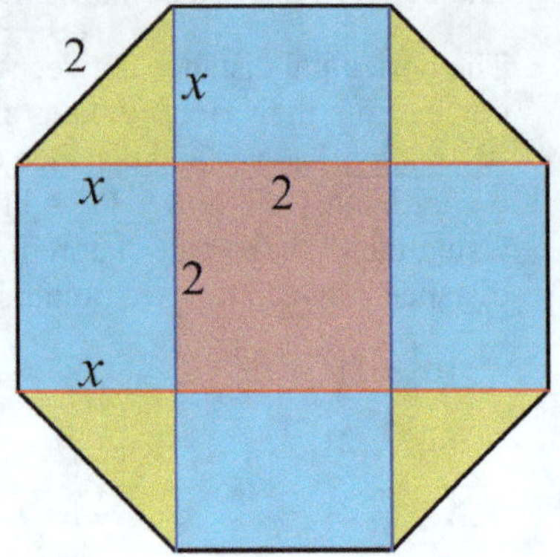

Puzzle corner. We divide the octagon into nine areas as in the image on the right. In the middle is a 2 by 2 square. Then there are four rectangles, with sides 2 and x units long. And in the corners we have four right triangles with base x and height x.

We can solve x using one of the corner triangles and the Pythagorean Theorem:

$$x^2 + x^2 = 2^2$$
$$2x^2 = 4$$
$$x^2 = 2$$
$$x = \sqrt{2}$$

Thus, the total area is: $4 + 4 \cdot 2 \cdot \sqrt{2} + 4 \cdot \sqrt{2} \cdot \sqrt{2} \div 2 = 4 + 8\sqrt{2} + 4 = 8 + 8\sqrt{2}$. which is about 19.3 square units.

Review, pp. 49-54

Page 49

1. a. 8 b. 13 c. 50 d. 0.9
   e. 6/10 = 3/5 f. −7 g. 5 h. 30

2. a. between 2 and 3 b. between 8 and 9 c. between 11 and 12 d. between 6 and 7

3. Here is one possible way the process could go. The intermediate guesses by the student(s) will vary. The table is just showing one example.

$\sqrt{13} \approx 3.6$

Low Guess	$(LG)^2$	$(HG)^2$	High Guess
3.6	12.96	13.3225	3.65
3.6	12.96	13.0321	3.61

4.

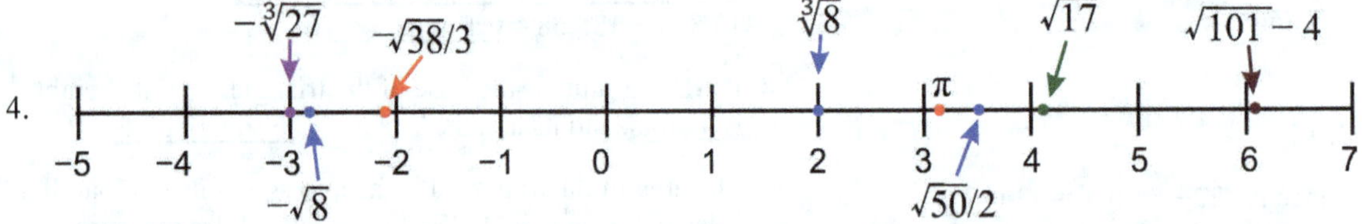

5.

a. $11 < \sqrt{150}$	b. $\sqrt{76} < 9$	c. $\sqrt{20} > 4$	d. $\sqrt[3]{10} > 2$
e. $4 < \pi + 1$	f. $\sqrt{85}/3 > 3$	g. $\sqrt{27} + 2 > 6$	h. $\sqrt{68} - 3 < 6$

Page 50

6. a. 12 b. −9 c. 40
   d. 8 e. 49 f. 20

7. a. 7 square units
   b. $4\sqrt{20}$ or $8\sqrt{5}$

8. a. rational; it is a decimal that terminates
   b. rational; the value is 50
   c. irrational; π is irrational, 5 is rational, and an irrational number times a rational is irrational
   d. irrational; 56 is not a perfect square, so $\sqrt{56}$ is irrational and so is its opposite
   e. rational; this is a ratio of two integers
   f. rational; the value is 1/6
   g. rational; this is a repeating decimal
   h. rational; this is a terminating decimal
   i. rational; this is a repeating decimal
   j. irrational; Seven is not a perfect square so $\sqrt{7}$ is irrational. Four is rational, and an irrational number times a rational number ($4\sqrt{7}$) is irrational.

Page 51

9. a.
$$\begin{aligned} 100x &= 61.6161\ldots \\ -\ x &= 0.6161\ldots \\ \hline 99x &= 61 \\ x &= \underline{61/99} \end{aligned}$$

b.
$$\begin{aligned} 10x &= 41.7777\ldots \\ -\ x &= 4.1777\ldots \\ \hline 9x &= 37.6 \\ x &= 37.6/9 = 376/90 = \underline{188/45} \end{aligned}$$

10. a. $x = \sqrt{147}$ or $x = -\sqrt{147}$
   b. $a = 13$ or $a = -13$
   c. $w = \sqrt[3]{0.36}$
   d. $x = \sqrt[3]{7}$
   e. $b = 5$
   f. $a = -2$

11. a.
$$\begin{aligned} y^2 + 18 &= 35 \\ y^2 &= 17 \\ y &= \sqrt{17} \approx 4.123 \\ \text{or } y &= -\sqrt{17} \approx -4.123 \end{aligned}$$

Check: $(\sqrt{17})^2 + 18 \stackrel{?}{=} 35$

$17 + 18 = 35$ ✓

b.
$$\begin{aligned} 0.6h^2 &= 4 \\ h^2 &= 4/0.6 = 40/6 = 20/3 \\ h &= \sqrt{20/3} \approx 2.582 \\ \text{or } h &= -\sqrt{20/3} \approx -2.582 \end{aligned}$$

Check: $0.6 \cdot (\sqrt{20/3})^2 \stackrel{?}{=} 4$

$0.6 \cdot (20/3) \stackrel{?}{=} 4$

$(6/10) \cdot (20/3) \stackrel{?}{=} 4$

$120/30 = 4$ ✓

Page 52

12. a.
$$\begin{aligned} 20^2 + 24^2 &\stackrel{?}{=} 30^2 \\ 400 + 576 &\stackrel{?}{=} 900 \\ 976 &> 900 \end{aligned}$$

No, they don't form a right triangle. (They would form an acute triangle.)

b.
$$\begin{aligned} 1^2 + 2.4^2 &\stackrel{?}{=} 2.6^2 \\ 1 + 5.76 &\stackrel{?}{=} 6.76 \\ 6.76 &= 6.76 \end{aligned}$$

Yes, they form a right triangle.

13. We can ignore the negative answers because a side cannot have a negative length.

a.
$$\begin{aligned} s^2 &= 3^2 + 5^2 \\ s^2 &= 9 + 25 \\ s^2 &= 34 \\ s &= \sqrt{34} \end{aligned}$$

b.
$$\begin{aligned} y^2 + 12^2 &= 14^2 \\ y^2 + 144 &= 196 \\ y^2 &= 52 \\ y &= \sqrt{52} \end{aligned}$$

14.
$$\begin{aligned} x^2 + 21.1^2 &= 22.5^2 \\ x^2 + 445.21 &= 506.25 \\ x^2 &= 61.04 \\ x &= \sqrt{61.04} \approx \underline{7.8\text{ m}} \end{aligned}$$

Page 53

15. Let x be the hypotenuse of this triangle.
$$\begin{aligned} (\sqrt{7})^2 + (\sqrt{8})^2 &= x^2 \\ 7 + 8 &= x^2 \\ x^2 &= 15 \\ x &= \sqrt{15} \end{aligned}$$

The hypotenuse is $\sqrt{15}$ units long.

Page 53

16. The pennant is an isosceles triangle.

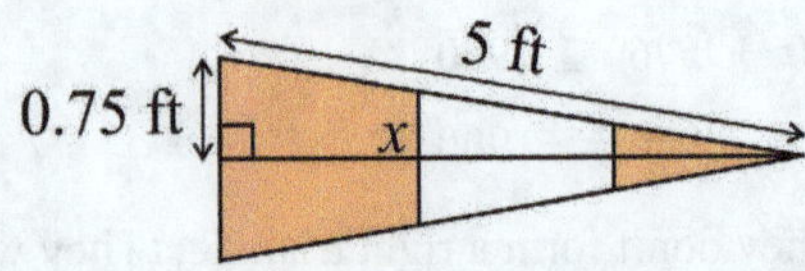

We calculate its altitude using the Pythagorean Theorem. From the right triangle in the image, we get:

$$0.75^2 + x^2 = 5^2$$
$$0.5625 + x^2 = 25$$
$$x^2 = 24.4375$$
$$x = \sqrt{24.4375} \approx 4.94343\ldots \text{ ft}$$

So the area is A = $bh/2 \approx 1.5$ ft · 4.94343 ft /2 = 3.70757 ft^2 ≈ <u>3.7 ft^2</u>.

17. The figure below has four right triangles, each with sides a, b and c. The sides of the outside square are $a + b$. The triangles enclose a square with sides c units long.

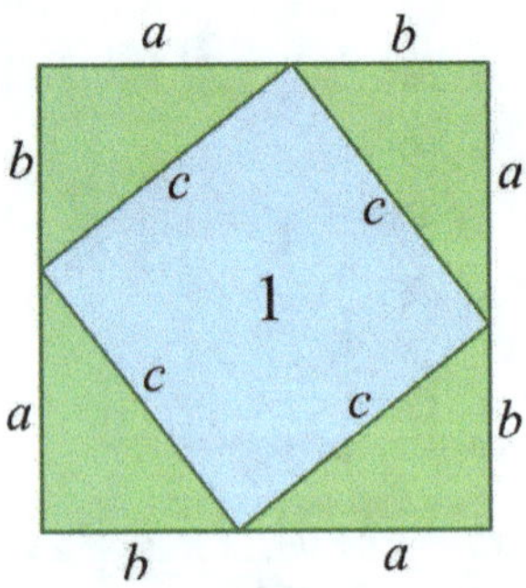

Below, the sides of the large square are still $a + b$, but but the four right triangles have been rearranged so that two smaller squares are formed, with sides a and b.

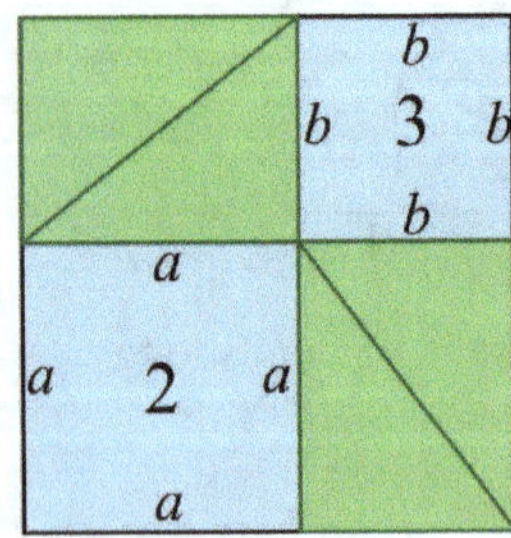

Since the areas of both large squares are equal, and the areas of the four right triangles are equal, it follows that the remaining (blue) areas are also equal. In other words, the area of square 1, which is c^2, equals the area of square 2 (which is a^2) plus the area of square 3 (which is b^2). In symbols, $c^2 = a^2 + b^2$. ☺

18. This is a trapezoid. Its area is the average length of the two parallel sides, times its altitude. The altitude is 5 units. The two parallel sides measure 5 and 9 units. So, the area is (5 + 9)/2 · 5 = 7 · 5 = 35 square units.

Since each unit is 2 ft, then one square unit in the image corresponds to 2 ft · 2 ft = 4 ft^2. Then, the area is 35 · 4 ft^2 = <u>140 ft^2</u>.

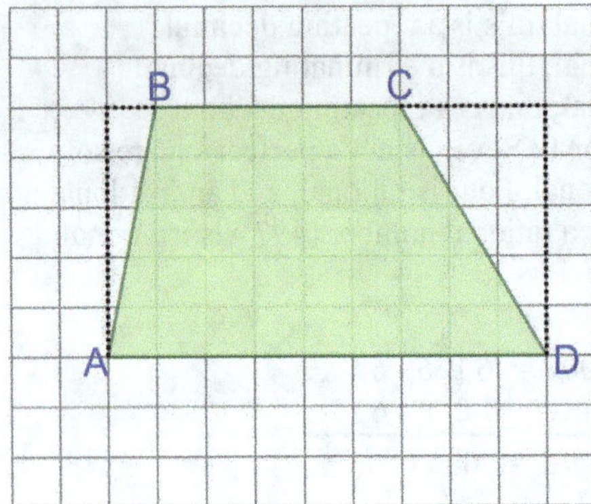

For the perimeter, we will find the side lengths AB and CD using the Pythagorean Theorem. The image shows with dashed lines the right triangles we will use.

For AB:
$$5^2 + 1^2 = AB^2$$
$$26 = AB^2$$
$$AB = \sqrt{26} \approx 5.099 \text{ units}$$

For CD:
$$3^2 + 5^2 = CD^2$$
$$34 = CD^2$$
$$CD = \sqrt{34} \approx 5.831 \text{ units}$$

The perimeter is then the sum of all four sides: P = 5.099 + 5 + 5.831 + 9 = 24.93 units. Since each unit is 2.0 feet, the perimeter is 49.86 feet ≈ <u>50 feet</u>.

Page 54

19. a. The horizontal distance between the points is $|11 - (-3)| = 14$ and the vertical distance is $|-9 - 4| = 13$.

$$x^2 = 14^2 + 13^2$$
$$x^2 = 196 + 169$$
$$x = \sqrt{365} \approx 19.1 \text{ units}$$

b. The horizontal distance between the points is $|70 - 42| = 28$ and the vertical distance is $|100 - (-15)| = 115$.

$$x^2 = 28^2 + 115^2$$
$$x^2 = 14{,}009$$
$$x = \sqrt{14{,}009} \approx 118.4 \text{ units}$$

Page 54

20. Please see the image below:

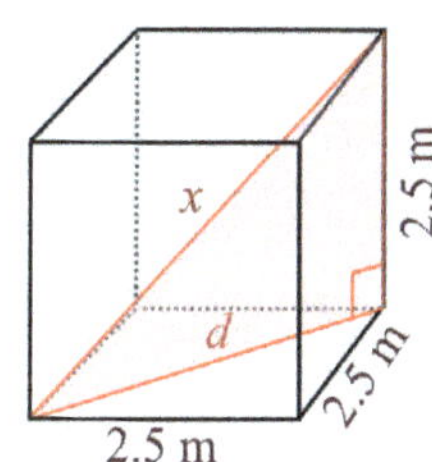

We will first solve for d using the right triangle on the bottom of the cube:

$$2.5^2 + 2.5^2 = d^2$$
$$12.5 = d^2$$
$$d = \sqrt{12.5}$$

Next, we will use the Pythagorean Theorem again in the pink triangle:

$$(\sqrt{12.5})^2 + 2.5^2 = x^2$$
$$12.5 + 6.25 = x^2$$
$$x = \sqrt{18.75}$$
$$x \approx \underline{4.3 \text{ m}}$$

21. Let x be the distance from B to C along 5th Avenue South.

Then:

$$370^2 + x^2 = 620^2$$
$$136{,}900 + x^2 = 384{,}400$$
$$x^2 = 247{,}500$$
$$x = \sqrt{247{,}500} \approx 497.49 \text{ m} \approx 500 \text{ m}$$

To go directly from A to C is 620 m, and the distance from A to B and then to C is 370 m + 500 m = 870 m. Therefore, to go directly from A to C is 870 m − 620 m = <u>250 m shorter</u> than to go from A to B and then to C.

Math Mammoth has a variety of resources to fit your needs. All are available as economical downloads, and most also as printed copies.

- **Math Mammoth Light Blue Series**
  A complete curriculum for grades 1-8. Each grade level includes two student worktexts (A and B), which contain all the instruction and exercises all in the same book, answer keys, tests, cumulative reviews, and a worksheet maker. International (all metric), Canadian, and South African versions are also available.
  https://www.MathMammoth.com/complete-curriculum
  https://www.MathMammoth.com/international/international
  https://www.MathMammoth.com/canada/
  https://www.MathMammoth.com/south_africa/

- **Math Mammoth Skills Review Workbooks**
  These workbooks are intended to be used alongside the Light Blue series full curriculum, and they provide additional review to the topics studied in the main curriculum, in a spiral manner.
  https://www.MathMammoth.com/skills_review_workbooks/

- **Math Mammoth Blue Series**
  Blue Series books are topical worktexts for grades 1-8, containing both instruction and exercises. They cover all elementary math topics from 1st through 8th grade and some for 8th grade. These books are not tied to grade levels, and are thus great for filling in gaps.
  https://www.MathMammoth.com/blue-series

- **Make It Real Learning**
  These activity workbooks concentrate on answering the question, "Where is math used in real life?" The series includes various workbooks for grades 3-12.
  https://www.MathMammoth.com/worksheets/mirl/

- **Review Workbooks**
  Workbooks for grades 1-8 that provide a comprehensive review of one grade level of math —for example, for review during school break or summer vacation.
  https://www.MathMammoth.com/review_workbooks/

Free gift!

- Receive over 350 free sample pages and worksheets from my books, plus other freebies:
  https://www.MathMammoth.com/free/

Lastly...

- Inspire4 is an inspirational website for the whole family I've been privileged to help with:
  https://www.inspire4.com

www.ingramcontent.com/pod-product-compliance
Lightning Source LLC
LaVergne TN
LVHW080926110826
845155LV00039B/220

* 9 7 8 1 9 5 4 3 5 8 7 4 4 *